THE ANNOTATED FLATLAND

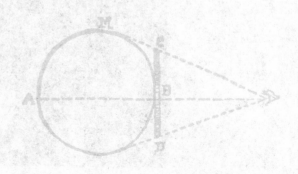

Edwin A. Abbott, the young headmaster.
Inscribed in pencil "Fradelle and Young, 283 Regent St. W."

THE ANNOTATED
FLATLAND

A ROMANCE OF MANY DIMENSIONS

By EDWIN A. ABBOTT

INTRODUCTION AND NOTES BY

IAN STEWART

BASIC
BOOKS

A MEMBER OF THE PERSEUS BOOKS GROUP   New York

Hardover edition first published in 2002 by Perseus Publishing,
A Member of the Perseus Books Group
Paperback edition first published in 2008 by Basic Books,
A Member of the Perseus Books Group

Cataloging-in-Publication Data is available from the Library of
Congress
Hardcover 0-7382-0541-9
Paperback 978-0-465-01123-0
LSC-C
10

Acknowledgments

Thanks to:

- The City of London School (especially Head Porter Barry Darling) for materials and access.
- Richard Baldwin, Cemeteries Manager, Camden, for telling me where to find Abbott's grave.
- Marianne Colloms of the Friends of Hampstead Cemetery, for sending me a photograph of Abbott's grave and alerting me to the existence of *The Good Grave Guide to Hampstead Cemetery, Fortune Green.*
- Terry Heard, former head of mathematics at the City of London School, for inside information about Abbott and his father.
- Jonathan Munn, graduate student at the University of Warwick, for discussions on four dimensions in theology.
- Bruce Westbury, mathematician at the University of Warwick, for information about polytopes and Alicia Stott Boole.

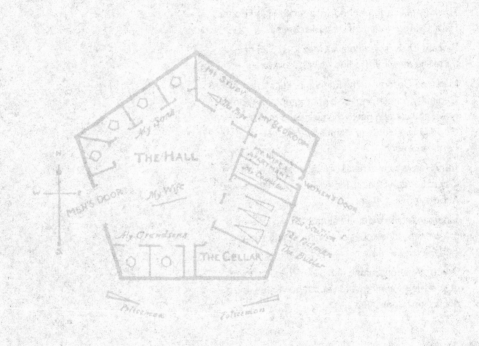

CONTENTS

PREFACE

What is *Flatland*, and why should it be annotated?

Flatland is a work of scientific fantasy written by the English clergyman and headmaster Edwin Abbott Abbott and published in 1884. It is
a charming, slightly pedestrian tale of imaginary beings: polygons who
live in the two-dimensional universe of the Euclidean plane. Just below
the surface, though, it is a biting satire on Victorian values — especially
as regards women and social status — and an accomplished and original piece of scientific popularization about the fourth dimension. And,
perhaps, an allegory of a spiritual journey.

It deserves to be annotated because, just as Euclid's plane is embedded in the surrounding richness of three-dimensional space, so *Flatland* is embedded in rich veins of history and science. Investigating these
surroundings has led me to such diverse quarters as *The Good Grave
Guide to Hampstead Cemetery,* phrenology, ancient Babylon, Karl Marx,
the suffragettes, the Indian Mutiny of 1857, the Gregorian calendar,
Mount Everest, the mathematician George Boole and his five remarkable
daughters, the Voynich manuscript, H. G. Wells's *The Time Machine,* the
"scientific romances" of Charles Hinton, spiritualism, and Mary Shelley's
Frankenstein.

I first read *Flatland* in 1963 as an undergraduate newly arrived at the University of Cambridge (England) to study mathematics. I enjoyed it, added it to my science fiction collection — it fits a broad definition of the genre — and forgot about it. Years later, I reread it, and the idea of a modern sequel began to form in my mind. I wasn't the first person to think of that, or to do it, but recent advances in science and mathematics made it easy for me to invent a new scenario. The result was *Flatterland*, whose genesis I have related in its own preface. While *Flatterland* was being readied for publication, my editor Amanda Cook at Perseus Books came up with the idea of a companion volume — a republication of the original *Flatland*, with added annotations.

I started with the idea that I would focus mainly on the mathematical concepts that *Flatland* uses or alludes to, so the writing ought to be simple and straightforward. But when I began looking into the life and times of its author, his associates, and the scientific and cultural influences that led up to the writing of Abbott's unique book, I was hooked. My amateur-historian investigations led into ever more fascinating byways of Victorian England and America, and I began to rediscover many things that are no doubt well known to Abbott scholars but are far from common currency.

At first, I was concerned that I might not be able to lay hands on the necessary material. But a glance at one of the more obvious and accessible sources, Abbott's entry in the *Dictionary of National Biography*, brought to light a curious coincidence. Ab-

bott's professional life revolved around the City of London School; he is its most famous headmaster, a post that he took up in 1865. Now, there exists in London an institution called Gresham College. It was founded in 1597 with a legacy from Sir Thomas Gresham (1518/19–1579), originator of Gresham's Law ("Bad money drives out good") and founder of the Royal Exchange in 1566–1568. Gresham was a philanthropist, and his will instructed the Mercer's Company (one of the livery companies created by King Richard II) and the city of London to "permit and suffer seven persons by them from time to time to be elected and appointed … sufficiently learned to read … seven lectures." Gresham College has no students — only the general public — and until recently it appointed seven professors, in astronomy, divinity, geometry, law, music, physics, and rhetoric. To these have been added an eighth: commerce.

Anyway, between 1994 and 1998 I was the Gresham Professor of Geometry. The first such was Henry Briggs (1561–1630, appointed in 1596), the inventor of "natural" logarithms. The others have included Isaac Barrow (1630–1677, appointed 1662), who recognized that differentiation and integration, the two basic operations of calculus, are mutually inverse; Robert Hooke (1635–1703, appointed 1664), who discovered the law of elasticity named after him, suggested that Jupiter rotates, and laid the early foundations of crystallography; and Karl Pearson (1857–1936, appointed 1890), one of the founders of statistics. The college is still funded by the Mercer's Company and the city of

London. The city also has a long-standing interest in the City of London School, and as a Gresham Professor I had lectured at its sister institution, the City of London School for Girls (founded in 1894). Thus I had an easy introduction to Abbott's professional home. The City of London School had been badly damaged in World War II and had moved to new premises; I wrote a letter asking whether it still had any Abbott documents, pictures, or other information. In response, Head Porter Barry Darling sent me a history of the school (*City of London School* by A.E. Douglas-Smith), which contained extensive information about Abbott, and invited me to visit and look through the school's archives.

A week later, I was ushered into a small, rather disorganized room lined with shelves and crammed to the ceiling with old books, magazines, photographs, and bound volumes of letters. On the top shelf, tucked away in one corner, was an almost complete collection of Abbott's books, including a rare first edition of *Flatland.* (I knew that a second, revised edition had followed hard on the heels of the first because the preface to that second edition says so. What had he changed? Now I could find out.) I went away with a stack of photocopies and with three framed photographs of Abbott at various stages of his career, lent to me for copying. I had his obituary in the school magazine, a review of *Flatland* in the same journal, samples from geometry texts used by the school in Abbott's day, extracts from his publications, and even a copy of his letter of resignation.

Of course, the Abbott scholars had been there before me, but even so, I felt like Sherlock Holmes hot on the trail of Moriarty.

Other sources now came into their own. I could surf the Net because I had some idea of what to look for. Entering "Flatland" into Yahoo turned up thousands of sites about off-road vehicles, but "Edwin Abbott Abbott" was much more helpful. An article by Thomas Banchoff (*the* leading expert on Abbott, currently working on a biography) explained the crucial connection to Charles Howard Hinton, whose wild but ingenious speculations about the fourth dimension undoubtedly inspired Abbott's fable. A conversation with a colleague, Bruce Westbury, in the Warwick University mathematics common room, put me on to the four-dimensional mathematics of Alicia Stott Boole. As a science fiction aficionado, I already knew that H.G. Wells had used four-dimensional geometry in *The Time Machine.* Now the Web turned up a brilliant historical survey by the science fiction author Stephen Baxter, and another by James Beichler, linking Wells to Hinton. Rudy Rucker's *The Fourth Dimension* opened up dozens of further leads ... and so it went.

What is the purpose of an annotated edition? Martin Gardener, in the classic among all such books, *The Annotated Alice: The Definitive Edition,* put it this way: "I see no reason why annotators should not use their notes for saying anything they please if they think it will be of interest, or at least amusing." Which is exactly my feeling. Accordingly, I pursued trails wherever they led and reported anything that seemed to fit the overall story. The most extreme case is a series of associations that links Abbott to Mary and Percy Bysshe Shelley, Lord Byron, Augusta Ada Lovelace, Charles Babbage, Sir Edward Ffrench Bromhead, George Boole, Mary Boole, Charles Howard Hinton, Alicia Stott Boole, and the Dutch mathematician Peiter Schoute — with a side branch to the science fiction writer H. G. Wells.

Something important emerges from such chains of connections: Victorian England was a tightly knit society. The intellectuals all knew each other socially, traded and stole each other's ideas, and married each other's sons and daughters. It was an exciting period of scientific and artistic discovery, for the staid and repressive attitudes of the Victorian era were sowing the seeds of their own destruction. Abbott knew many of these people — most of them more colorful than he was — and they influenced his thinking in profound ways. It's been fun ferreting out their stories. For example, along the way I discovered that I once held the same job as Abbott's mathematics teacher — but 146 years later.

As a strictly amateur historian, I *know* that I will have made some mistakes, misinterpreted some events, or left out some vital items of information that are well known to all the experts. This happens with any book; it is virtually impossible to track down *all* the relevant documentation, all the names, all the dates. (I've been moderately obsessive about giving dates for almost everything and everybody — except for minor figures — because the *timing* is so crucial in this kind of investigation. When I am not sure of a date, I've either followed the date with a question mark or omitted it.) Hence I invite anyone who has constructive criticisms, useful observations, wild theories, or new information to e-mail them to Flatland@JoatEnterprises.co.uk. I can't promise you a reply, though I'll do my best, but I do promise that I'll take note of anything I think is interesting. And when it is time to prepare a new edition, I'll make the necessary changes.

I also promise that *nearly* everything I say is true — or, if it's an opinion, plausible. I've tried to do my historical and scientific homework. I hope you'll come to agree with me that there is so much more to *Flatland* than meets the eye, even if it is a world of only two dimensions.

Coventry, May 2001

INTRODUCTION

In mathematical and scientific circles, Edwin Abbott Abbott is known for one thing and one thing only: his mathematical fantasy *Flatland.* To his contemporaries, however, he was renowned as a teacher, writer, theologian, Shakespeare scholar, and classicist. His entry in the *Dictionary of National Biography* occupies more than two double-column pages, yet *Flatland* is not mentioned.

Abbott was born in London on 20 December 1838. His unusual double-barreled name arose in part because his father Edwin Abbott (1808–1882) married a first cousin, Jane Abbott. Abbott (from now on this name will refer to Edwin Abbott Abbott) was educated at the City of London School, starting in 1850, and he showed early talent. There he met another pupil, Howard Candler, who became a lifelong friend. He studied mathematics under the brilliant but eccentric Robert Pitt Edkins (?–1854). (In 1848 Edkins was appointed Gresham Professor of Geometry, which underlines the close connections between Gresham College and the City of London School and explains in what sense I have had the same job as Abbott's mathematics teacher.) In 1857 Abbott won a scholarship to St. John's College at the University of Cambridge to study classics. There he made brilliant progress: He was senior classic and senior chancellor's medalist in 1861 and was elected a fellow of the college in 1862 (and an honorary fellow in 1912). In 1863 he married Mary Elizabeth Rangeley, daughter of the Derbyshire landowner Henry Rangeley, resigning his college fellowship to do so. He was ordained a deacon in the Anglican Church in 1862 and became a priest in 1863.

Although Abbott had the intellectual ability to become a first-class university academic, his interests lay elsewhere: He was an enthusiastic and capable educator, and he spent most of his professional life as a teacher. His first teaching post was as an assistant master at King Edward's School, Birmingham, in 1862. He moved to Clifton College in 1864, but in 1865 he returned to his old school, the City of London School, as headmaster. At the time he was only 26 years old and had the difficult task of presiding over a substantial number of much older teachers who had *taught* him when he was a pupil, some of whom had also applied for the headship. He won their respect and became the school's best-known headmaster. He remained at the school, as head, until his retirement in 1889. His reputation was such that he was repeatedly urged to accept the headmasterships of such major "public" schools as Rugby, Marlborough, and Wellington. (England's public schools were, and still are, privately owned, and they cater mainly to the children of the upper classes. The name, which arose in the eighteenth century, indicated that they drew their pupils from the country as a whole, not just from their local area.) Balliol College, Oxford, tried to secure him as a theology lecturer. "I am so bothered and bullied," Abbott wrote to his close friend Howard Candler; refusing all offers, he remained at the City of London School.

Abbott was an outstanding teacher. He was a small man, but he used his piercing eyes and sonorous voice to good effect. R.S. Conway, in an obituary in the *Manchester Guardian*, wrote,

One of Abbott's commonest methods of dealing with a muddled answer was to turn to another boy and bid him repeat the explanation given; and when it appeared that the muddle had produced a still worse muddle in the second boy's mind, Abbott would look back to the first boy and say "You see how far you have made him understand." If a clever boy at the top [of the class] gave an answer which was correct but too brief or too technically worded to be readily understood, he would say "Yes, yes, all very well; but think of the boy down there at the bottom." When Jones made a mistake in the construing, Abbott would turn sharply and say "Correct that, Brown," and when, as often happened, Brown produced a merely satisfactory version which did not make clear where Jones had gone wrong, Abbott would say: "No, no, what was his mistake? Tell him his mistake."

Abbott was a tireless worker, typically getting up at 5 A.M. so that he could write — on Latin, English, theology, and other topics — before starting his day at the school. Under his direction, the school became one of the best in the country. Although his own background was classics (Latin and Greek), he made sure that the curriculum included substantial amounts of English literature, especially Shakespeare. When H.C. Beeching credited Mortimer with introducing English literature to the school, Abbott set him straight:

[D]on't say that Dr. Mortimer introduced the study of English Literature. For he did not. I was some six or seven years in the School and never was taught a word of it....The study of English Literature was introduced by ME, with Seeley's aid and impulse.

The boys had to study one Greek and one Shakespearean play every term. On "Beaufoy Day," when the prizes were awarded, it became traditional to have recitations from Shakespeare; under Abbott these developed into dialogues and eventually the performance of entire scenes. Unusually for that period, Abbott also emphasized the importance of science, making chemistry a compulsory subject for all pupils. Mathematics, too, was given prominence and was taught to a high level. Abbott taught classes in comparative philology — the relationships among different languages — and the really enthusiastic pupils got a dose of Sanskrit. In 1878 Abbott even introduced shorthand as an optional subject, after the School Committee had seen a shorthand Bible written by E.H. Hone.

Abbott was an effective administrator, and he built the school up from one of ordinary stature to one of considerable eminence. He reduced class sizes, which had often reached seventy pupils per teacher when he first arrived. He hired high-quality teachers. In the *Dictionary of National Biography,* his former pupil Lewis Richard Farnell (1856–

Edwin A. Abbott in his prime.
Inscribed "W. & D. Downey, 57 & 61, Ebury St., London."

1934) says that Abbott "had the mark of the spiritual leader in that he could impart to others something of the 'virtue' that was in him. He was aflame with intellectual energy: without driving or over-taxing his pupils he made intellectual effort a kind of religion for them."

Abbott was a religious reformer, too: He was a passionate and articulate member of the Broad Church, which was opposed to the mystical language and dogma of both the High and Low branches of the Church of England. The Broad

Church instead espoused social democracy with a Christian slant. He was an effective preacher with a simple, clear delivery. Cambridge University made him Hulsean lecturer in 1876; Oxford responded by making him select preacher the year after. His sermons for both appointments were later published (see "Bibliography of Edwin Abbott Abbott" at the back of this book).

By the 1870s it was becoming clear that the existing school was too small and outdated, and in December 1874, Abbott wrote to his friend Candler: "Next year I shall agitate for the removal of the School." It proved to be the start of a lengthy and debilitating struggle, which ultimately led to Abbott's premature resignation. The classes were seriously overcrowded, and Abbott fulfilled his promise to Candler in February 1875, when he presented a report to the School Committee on "the opportunities of recreation and amusement provided and on the moral and physical disadvantages arising from defective sanitary arrangements and the want of a Playground." Thus began a seven-year battle. In 1878, after years of wrangling with the School Committee, Common Council of the City of London, and the City Lands Committee, the city architect drew up plans for a new school building on the Victoria Embankment site. Finally, on 14 November 1878, the Common Council made the decision to go ahead. It was not an easy decision: The new school would cost £200,000, a huge sum at that time. In 1879, the School Committee chose a design for the building, and Abbott wrote,

At last! After about five years of reporting on my part, persuading, experimenting, conquering, retreating, being promised, being cheated, to succeed at last! To have a good school, with good rooms and good appliances, on such a site, to last as long as London lasts: the more I think of it the more I rejoice.

In 1880 Abbott announced publicly that he expected to be in the new school within a year. Not so. There was a change of plans. Part of the site was allocated to Sion School, with a compensatory addition of land elsewhere, and this entailed drawing up entirely new architects' plans. In 1880 Abbott showed signs that his mind was on other things and that once the boys were settled in the new school, he might call it a day. In 1882 he wrote, "If I leave the School next October I shall have led the School into the Promised Land and may fairly give place to a younger Joshua." In fact, he stayed on for seven more years. The new school was opened in late 1882, and staff and pupils were transferred to it on 23 January 1883. The old school was sold for £60,000.

During this tiresome period, it was becoming ever more clear to Abbott that his *real* calling was as an author. He had started publishing books in 1871, six years into his headship at the City of London School. Several of his books were used by the school as texts. In all, he published over fifty books (see the Bibliography), which fell into three categories: school texts, literary scholarship, and theology. It is the exception that is of most interest to us here. Unique among all of his writings — indeed, nothing really like it can be found anywhere else in

English literature — is the mathematical fantasy *Flatland.*

Flatland: A Romance of Many Dimensions was probably written over the summer of 1884. It is quite unlike anything else that Abbott wrote, and it appeared under the pseudonym of A. Square. (Several of Abbott's other works were also first published pseudonymously: *Philochristus, Onesimus,* and *The Kernel and the Husk.*) In October 1884, Abbott distributed a draft to a select group of friends and acquaintances. A first edition of 1,000 copies was published in November of the same year by J. R. Seeley of London (address: 46, 47, and 48 Essex Street, the Strand). It was printed by R. Clay, Sons, and Taylor of Bread Street Hill, and the price was half a crown. In December, only a month later, a revised second edition appeared; it contained a new preface by the author.

Flatland is not a lengthy book. It is 102 pages long (100 in the first edition) with a further 10 pages (8 in the first edition) of "forematter" — title page, dedication, introduction, preface, contents. The setting for the tale is a two-dimensional world, an infinite Euclidean plane: Flatland. Flatland is inhabited by intelligent creatures shaped like geometric figures — Lines, Triangles, Squares, Hexagons, Circles. The book is divided into two parts of equal length. The first part, which describes Flatland's customs, society, and history, is mainly satire aimed at Victorian social assumptions — especially the subservient role of women and the rigid class-ridden hierarchy of the men, in which advancement is dependent on geometric regularity. It ends with a plea for a reevaluation of women's education — ostensibly in Flatland, but aimed at Victorian England. The real aim of the second part of the book is to introduce the concept of the fourth dimension, which it does by analogy. The difficulties that a two-dimensional being experiences in comprehending the *third* dimension are used to help Victorians living in a three-dimensional space accustom themselves to the radical but enormously popular idea of a fourth dimension. The vehicle for the explanation of these ideas is to recount the adventures of a character named A. Square, who explores spaces of various dimensions: Flatland itself (two-dimensional, a perfect Euclidean plane), Lineland (one dimension), Pointland (no dimensions), and Spaceland (three dimensions).

In his book *The Fourth Dimension,* Rudy Rucker points out that there is a third, less obvious aspect to *Flatland* — that of a spiritual journey, which would fit with Abbott's theological leanings. "A Square's trip into higher dimensions is a perfect metaphor for the mystic's experience of a higher reality." In his introduction to the Princeton Science Library's edition of *Flatland,* Banchoff makes a related point:

The narrative style of *Flatland* is somewhat different from some of the more familiar reports of visits to exotic lands, since the story is told not by the visitor but the person visited. It is as if the story of Gulliver were told by the Mayor of Lilliput or the adventures of Alice by the White Rabbit. . . . The reader has to stay with the story all the way through to appreciate the change that takes place in the storyteller. A similar thing happens with the narrator of another book written in exactly the same year, *Huckleberry Finn.*

There are difficulties in establishing the precise reaction of Victorian readers to *Flatland*'s unusual setting and style. In particular, an extensive archive of Abbott's letters and personal documents, borrowed from the City of London School during the preparation of a book on its history, mysteriously disappeared about sixty years ago. To most Victorians, Abbott's important works were his theological and scholarly ones. The history of the school makes only passing mention of *Flatland,* and Abbott's obituaries and his entry in the *Dictionary of National Biography* do not mention it at all. We gain one useful insight into *Flatland*'s reception from the preface to the second edition, where Abbott includes an explicit (and slightly pained) reaction to criticisms of the book's treatment of women. It seems likely that this feature of the book had been the subject of heated comment, though again no documentary evidence seems to have survived. We do know that Abbott was in regular contact with several female intellectuals, among them the novelist George Eliot (Mary Ann Evans, 1819–1880) and the educators Dorothea Beale (1831–1906) and Frances Mary Buss (1827–1894). They probably understood Abbott's intent, which was satirical: By making Flatland men treat their women with undisguised contempt, he was pointing out how common this attitude was in Victorian society. However, some of Abbott's readers must have misunderstood the irony — enough of them to prompt

his rapid clarification in the second edition. In the same gentle manner that characterized his teaching, Abbott phrased this clarification as a change of attitude by A. Square, who "has himself modified his own personal views."

There is a rather bland review of *Flatland* on pages 217–221 of the *City of London School Magazine,* vol. 8, 1884 (also printed by Clay, Sons, and Taylor). The review, which is unsigned, begins,

We have strong reasons for believing that the author of the above is not unknown to most of our readers: that this is not the first or the most philosophical production of his pen: and, what is more to the point, that the name of A SQUARE will be found in the Mathematical Tripos list for the year 186– by anyone who will consult the Cambridge Calendar for that purpose.

The Mathematical Tripos is a series of examinations for the mathematics degree at Cambridge University, whereas Abbott studied classics. However, this reference is not an error. In *Mathematical Visions: The Pursuit of Geometry in Victorian England,* Joan L. Richards says that "at Cambridge, even if Greek and Latin were his major interest, a student could not take the classics examination without first passing the mathematical tripos." The review then proceeds with a rather plodding summary of the plot, and not much else, until the final paragraph:

If, however, at any time the author is fortunate enough, in his capacity of a solid, to receive a revelation from the universe next above us in the continuous scale, we entreat him to lose no time in transferring to paper and transmitting to posterity his adventures in that region also. We at least shall be grateful to him.

In 1885 there was an American edition, and a Dutch one followed in 1886. The book was reprinted several times in the United States, which indicates its continuing popularity there, but it seems to have disappeared from the British scene until 1926, when the second edition was reprinted by Basil Blackwell with an added introduction by William Garnett. This is the edition I have chosen to annotate because Garnett's introduction develops the scientific background and shows how much progress had been made in the forty-two years following the first edition. Numerous editions have been published since, with introductions by Ray Bradbury, Karen Feiden, Isaac Asimov, Alexander Keewatin Dewdney, Banesh Hoffman, and Thomas Banchoff. There exist translations into at least nine languages, and at least twelve English-language editions are currently in print — partly, it must be said, because the lapsed copyright makes royalty payments unnecessary.

What led a respectable headmaster at a leading school and a priest in the Anglican Church to write an intellectually lightweight work of scientific fantasy? The precise reasons are not known, but there is a key figure in the events that both preceded and followed the writing of *Flatland*: Charles Howard Hinton (1853–1907). In 1990 Thomas Banchoff, the world's leading Abbott scholar, pointed out in *Interdisciplinary Science Reviews* that Hinton lies at the center of a web of intellectual, mathematical,

and social influences. These involve, for example, the mathematician George Boole, along with his family, and Herbert George Wells (author of *The Time Machine* in 1895). We explore some of these connections further in the annotations, but let me give a brief outline now.

First, the intellectual link. The fourth dimension was very much "in the air" in the late 1800s. The interest began among scientists and mathematicians, but their excitement transmitted itself to the general public. Hinton, one of the prime agents of that transmission, was well suited to the task: He liked writing for the general public, and he was a talented mathematician who had no trouble with the technical aspects of four-dimensional geometry. While in school at Rugby, he became interested in the works of George Boole, inventor of Boolean algebra, the logical basis of computer science. (Later, Hinton married Boole's eldest daughter Mary.) An 1869 letter written by Hinton's father says that at Rugby his son became interested "in studying geometry as an exercise of direct perception." In 1871 Hinton proceeded to Balliol College, Oxford University, where he took top honors in mathematics. We know from his books that in 1904 he had an excellent understanding of non-Euclidean geometry. Hinton could hardly have failed to become aware of a rash of astonishing new developments in the mathematics of four or more dimensions (see "The Fourth Dimension in Mathematics," which follows the annotated text of *Flatland* in this book), and these caught his imagination. He was also something of a mystic, and both then and later he related the fourth dimension to pseudoscientific topics that ranged from ghosts to the afterlife. (A ghost can easily appear from, and disappear along, a fourth dimension, for instance.) In this regard he was influenced by the unorthodox views of his father James Hinton (1822–1875), an ear surgeon. Hinton senior worked with Havelock Ellis (1859–1939), who outraged Victorian sensibilities with his frank studies of human sexual behavior. The elder Hinton became something of a libertine, advocating free love and polygamy and eventually heading a cult. He once said "Christ was the Saviour of Men, but I am the saviour of women, and I don't envy Him a bit."

The younger Hinton also had an eventful private life and seems to have taken his father's teachings a little too much to heart: In 1886 he was forced to flee to Japan, having been put on trial for bigamy at the Old Bailey. While married to Mary Boole — with whom he fathered four sons: Eric, George, Sebastian, and William — he had also married a certain Maud Weldon. He had spent a week with her in a hotel near King's Cross and, it was alleged, was the father of her twin children. He was in prison for only one day (some sources say three); Mary refused to pursue the matter. In Japan Hinton worked as a teacher in a Yokohama middle school, but by 1893 things were looking up, and he became a mathematics instructor at Princeton University. Here he invented a baseball pitching machine that propelled the balls via a charge of gunpowder. It was used for team practice for a while, but it proved a little too ferocious, and after several accidents it

was abandoned. Hinton was fired, whereupon he moved to the University of Minnesota.

His continuing efforts to promote the concept of the fourth dimension in the United States were wildly successful, and the topic appeared in popular magazines such as *Harper's Weekly, McClure's,* and *Science.* In 1900 he changed jobs, moving to the Naval Observatory in Washington, D.C. While there he wrote several articles on the fourth dimension (see "Bibliography of Charles Howard Hinton" at the back of this book), including one for *Harper's Monthly Magazine* in 1904. Around that time Hinton changed jobs again, this time joining the patent office in the same city. (In a sense, Hinton's career finished where Einstein's started, for Einstein was a clerk in the patent office in Bern from 1902 to 1908. Their stints even overlapped, temporally if not geographically.) Hinton died suddenly in 1907. A newspaper report, with the headline SCIENTIST DROPS DEAD, says that he was attending the annual banquet of the Washington, D.C. Society of Philanthropic Enquiry. Having just complied with the toastmaster's request for a toast to "female philosophers," Hinton collapsed and died on the spot of a cerebral hemorrhage. In 1909 *Scientific American* offered a $500 prize for "the best popular explanation of the Fourth Dimension" and was deluged with entries. Many commented favorably on Hinton's work.

Hinton's influence on *Flatland* appears to have originated in 1880, when he published an article entitled "What Is the Fourth Dimension?" in the *Dublin University Magazine.* In 1881 it was reprinted in the Cheltenham Ladies' College magazine (whose title, in some sources, is given as *Cheltenham Ladies' Gazette*), and in 1884 it appeared as a pamphlet to which the publisher had added the subtitle "Ghosts Explained." In it we find the following passage:

Suppose a being confined to a plane superficies [that is, surface], and throughout all the range of its experience never to have moved up or down, but simply to have kept to this one plane. Suppose, that is, some figure, such as a circle or rectangle, to be endowed with the power of perception; such a being if it moves in the plane superficies in which it is drawn, will move in a multitude of directions; but, however varied they may seem to be, those directions will all be compounded of two, at right angles to each other.

Here we find several essential plot elements of *Flatland*: a plane world, conscious creatures shaped like circles and polygons, and the geometric limitations of such creatures. A few paragraphs earlier, Hinton describes an even simpler world, along with its most striking limitation:

In order to obtain an adequate conception of what this limitation [to three dimensions of space] is, it is necessary to first imagine beings existing in a space more limited than that in which we move. Thus we may conceive of a being who has been throughout all the range of his experience confined to a single straight line. The whole of space would be to him but the extension in both directions of the straight line to an infinite distance. It is evident that two such creatures could never pass one another.

This, in all but name, is Abbott's Lineland, which plays a key role in the second half of *Flatland* by giving A. Square an analogy that helps him understand his own relation to three-dimensional space. Hinton makes it explicit that his two low-dimensional "worlds" are the first of a series when he links them with the sentence "to go a step higher in the domain of a conceivable existence." He also lays the groundwork for Abbott's entire cast of characters: "In a plane, there is a possibility of an infinite variety of shapes, and the being we have supposed could come into contact with an indefinite number of other beings. He would not be limited, as in the case of a creature on a straight line, to one only on each side of him."

The similarities do not stop there. Hinton points out that

If … we suppose such a being to be inside a square, the only way out that he could conceive would be through one of the sides of that square. If the sides were impenetrable, he would be a fast prisoner, and would have no way out.... Now, it would be possible to take up such a being from the inside of the square, and to set him down outside it. A being to whom this had happened would find himself outside the place he had been confined in, and he would not have passed through any of the boundaries by which he was shut in.

This feature of a plane world has several echoes in the plot of *Flatland*, including A. Square's dramatic abduction by the Sphere. Moreover, Hinton introduces these ideas in order to set up the same "dimensional analogy" that informs the storyline of

Flatland: A. Square is to space of three dimensions as a Victorian human is to space of four dimensions. Hinton then introduces a four-dimensional figure that he calls a four-square and we now call a hypercube: It is the four-dimensional analogue of a one-dimensional line segment, a two-dimensional square, and a three-dimensional cube. He points out numerical patterns in the numbers of corners of these shapes: 2, 4, 8, and 16, respectively — the first four powers of 2. Abbott does the same in the second half of *Flatland,* using the pattern as an argument to convince A. Square of the reality of three dimensions, and later invoking it again as A. Square tries to convince the Sphere of the reality of four dimensions.

Moving toward his main objective, Hinton describes some of the weird properties of four-dimensional space:

A being in three dimensions, looking down on a square, sees each part of it extended before him, and can touch each part without having to pass through the surrounding parts....So a being in four dimensions could look at and touch every point of a solid figure.

Abbott repeatedly includes references to creatures from a higher dimension being able to "see the insides" of lower-dimensional beings.

Hinton ends with two sections that have no real parallel in *Flatland:* He discusses the physics of four-dimensional space and evidence for its possible existence. These topics are beyond the scope of Abbott's story and inconsistent with its stance, but the similarities between Hinton's 1880 article and Abbott's 1884 book are far too great to be coincidence. However, although Hinton's influence on Abbott seems evident, there is no clear record that the two men ever met, either prior to the crucial summer of 1884, when *Flatland* was almost certainly being written, or afterwards. On the other hand, the circumstantial evidence that they probably did meet — or that, at the very least, Abbott was strongly influenced by Hinton's ideas — is considerable, as I will now explain.

Abbott was very much interested in the education of women, which in his day was extremely limited and unambitious — something that he wanted to change. In the 1880s he became involved in the promotion and reform of women's education. These activities led him to interact with Dorothea Buss, who at that time was headmistress of Cheltenham Ladies' College. In 1875 Hinton was appointed to teach at the college, so Abbott would almost certainly have met Hinton there. Also, Abbott could easily have come across an early draft of Hinton's essay, which was circulating in the college in 1884, and very probably earlier.

If not, another opportunity soon arose because in the same year Hinton moved to Uppingham School to become science master. The mathematics master at Uppingham School was Abbott's best friend Candler. Abbott and Candler first met when they were pupils at the City of London School; they went to Cambridge University together, and for twenty-five years — all the time Candler was at Uppingham — they wrote to each other every week. (Abbott's half of this correspondence was still in existence in 1939, when it supplied information for a history of the City of London School, but when the author of that history died, a thorough search revealed no trace of the letters. It seems unlikely that they were destroyed, and Abbott scholars fervently hope that they will turn up again.) Abbott and his wife often visited Uppingham to meet the Candlers; moreover, Edward Thring, the headmaster of Uppingham School, was the founder of the Head Master's Conference, of which Abbott was the secretary. It was Thring who undertook the task of sacking Hinton for bigamy.

Abbott was probably influenced by other thinkers, as well, as he developed the ideas in *Flatland*. His position in society was a respectable one, and he met and corresponded with many of the great figures of the time. One of his pupils was Herbert Henry Asquith (1852–1928), who went on to become prime minister. Abbott corresponded with George Eliot, and he was invited to her home in 1871, the same day that the physicist John Tyndall (1820–1893) went there for dinner. Tyndall showed experimentally that the sky is blue because sunlight is scattered by molecules in the air, an explanation later given a theoretical basis by Lord Rayleigh. In 1881 he drove the final nail into the coffin of the theory that life could be "spontaneously generated." From Tyndall, Abbott may have heard about the work of Hermann von Helmholtz (1821–1894), who wrote about non-Euclidean geometry. In public lectures, Helmholtz explained the new geometry in terms of an imaginary two-dimensional creature living on some mathematical surface, unaware of

EDWIN A. ABBOTT, MARCH 1914.

anything else and trying to find out about the geometry of its universe using only intrinsic features.

Several other authors have taken up Abbott's ideas and developed them, or ideas like them. The first to do so was Hinton, in his 1907 *An Episode of Flatland: How a Plane Folk Discovered the Third Dimension*. Hinton's "Flatland" is not the same as Abbott's, although it is two-dimensional. Instead of allowing his creatures to occupy the whole of the plane (Flatland "space"), Hinton confines them to the surface of a circular planet, Astria, floating in a planar (actually, slightly curved) universe. This offers a much closer analogy to our own situation, but it complicates the narrative — for example, if two beings meet going opposite ways, one must climb over the other. Astrians have a complicated physiology, but Hinton chose to draw them as right-angled triangles equipped with two arms, two legs, one eye, one nose, and one mouth. All males are born facing east, all females facing west. Barring intervention from the third dimension, they must remain that way throughout their lives. The plot is socialist, advocating the replacement of individual wealth by social planning. Astria is threatened by another planet, Ardaea, which will approach so closely that Astria's orbit will change into an eccentric ellipse, causing enormous climatic fluctuations. The world is saved by Hugh Farmer, a character who bears a remarkable resemblance to Hinton himself, and as such is, of course, a believer in the third dimension. He is sure that all Astrian objects have a slight extension into this new dimension, giving them a small but nonzero thickness. There may therefore exist beings that live outside the Astrian plane. The hero advocates a mass exercise in telekinesis — the (hypothetical) ability to move objects by thought alone—to latch on to such an external being and divert Astria away from the inbound rogue planet.

The use of the name Flatland by both authors seems not to have caused any personal animosity: Certainly, each later refers to the work of the other in (grudgingly) favorable terms. Abbott's 1887 *The Kernel and the Husk* mentions "a very able and original work by Mr. C.H. Hinton" about "a being

of Four Dimensions," and Hinton's *Scientific Romances* of 1888 refers to "that ingenious work *Flatland*." However, Abbott goes on to say that the ability to conceive of a space of four dimensions renders us "not … a whit better morally or spiritually," and Hinton complains that "the physical conditions of life on the plane have not been [Abbott's] main object." All this leads Banchoff to suggest that Hinton and Abbott "saw their efforts as complementary rather than competitive," which seems a fair assessment.

Hinton had some strange ideas about the fourth dimension, which were rooted in spiritualism and the paranormal. However, he also thought carefully about the science of four dimensions — and two. For example, he explains that the force of gravity in the Astrians' universe varies inversely as the distance, not inversely as its square. This is a natural choice in two dimensions: It ensures that the total amount of gravitational force exerted on a circle is the same, no matter how large the circle is, because the circumference of the circle is proportional to its radius. In three dimensions, the surface area of a sphere is proportional to the *square* of its radius (the area is $4\pi r^2$, where r is the radius), so an inverse-square law is required to achieve the analogous goal. Hinton does not examine the stability of planetary orbits with such a law of gravitation, but according to *Principles of Mechanics* by John L. Synge and Byron A. Griffith, a circular orbit subject to inverse nth-power gravitational attraction is stable if and only if $n < 3$, which is the case in Hinton's universe — and our own.

Of course, contributions to the Flatland canon have continued into our own time. In 1957 the Dutch physicist Dionys Burger wrote *Bolland* (translated in 1965 as *Sphereland*). He developed Flatland into a vehicle for explaining Einstein's ideas on the curvature of space-time and along the way, added many fascinating new details about the Flatlanders' lifestyle. Rudy Rucker's short story "Message Found in a Copy of *Flatland*," from his book *The 57th Franz Kafka*, provides an outrageous fictional explanation of how a staid clergyman came to write such an uncharacteristic book. The story, which is reprinted in his 1987 anthology *Mathenauts*, begins with an enigmatic note:

Robert—
   Flatland is in the basement of our shop. Come back at closing time and I will show it to you. Please bring one hundred pounds. My father is ill.
                                                    Deela.

In *The Shape of Space* (1985), Jeffrey R. Weeks interwove further fragmentary tales of Flatland with a mathematics text on the topology of surfaces and three-dimensional spaces. Rudy Rucker's *The Fourth Dimension* (1985) extended the Flatland canon in several new directions. There is also Alexander Keewatin "Kee" Dewdney's *The Planiverse* (1984), described below, but narratively this connects more closely to Hinton than to Abbott. Finally (for the moment), my own *Flatterland* (2001) brought the satire, the mathematics, and the physics into the nascent twenty-first century. It re-spects Abbott's narrative but makes no attempt to be consistent with any other sequel.

Anyone who thinks seriously about a world of two dimensions soon realizes that many things we take for granted in three dimensions don't work in two. There are, for example, no knots; telephone lines cannot cross without intersecting; organisms cannot have (conventional) digestive tracts or they would fall apart; and if you drive a nail through a plank of wood, the wood splits into two separate pieces. What would a Flatland bookcase look like? And what of gravity, chemistry, and atomic physics? In 1977 Dewdney began some serious thinking about two-dimensional physics and corresponded with various colleagues: The result was a fascinating set of duplicated notes, *Two-Dimensional Science and Technology* (1979, revised 1980). An article by Martin Gardner in *Scientific American* drew attention to the project, and the ideas that flooded in were compiled as *A Symposium on Two-Dimensional Science and Technology* (1981). By 1984 this had given rise to *The Planiverse*, a science fiction story set on the two-dimensional world of Arde. Dewdney offers lengthy and carefully thought-out descriptions of how two-dimensional physics and technology might actually work. For example, there is a design for an internal combustion engine. The appendix to *The Planiverse* recounts the history of the original project.

Abbott devoted his lengthy retirement to his writing, mainly a long series of books on theology under the collective title *Diatessarica*, which means, roughly, "matters pertaining to the harmony of the

ABBOTT'S FORMER HOUSE AT WELLSIDE, WELL WALK, HAMPSTEAD.

four gospels." On 12 October 1926, after a seven-year illness, he died at his home of Wellside, Well Walk, Hampstead — which at that time was a suburb on the northern outskirts of London, near a famous area of open country known as Hampstead Heath. The Heath still exists, and so does the house. Take the Northern Line of the London Underground to Hampstead station, exit the station to the left, and take a sharp left down Flask Walk, which runs into Well Walk. Wellside is the next to last house on the right, just before the junction with Heath Road; its name is clearly visible on one wall (above). The

house bears no plaque commemorative of its inhabitant — unlike 14 Well Walk, a few dozen yards away, where Marie Stopes (1880–1958), controversial founder of the United Kingdom's first birth control clinic, lived prior to 1916.

Abbott's funeral was at Christ Church, Hampstead, and he is buried in the southwest corner of Hampstead Cemetery. To visit his grave, take the Jubilee Line of the London Underground to West Hampstead station, exit the station to the right, and walk for about fifteen minutes up West End Lane and Fortune Green Road. The entrance to the cemetery is on the left, just before the junction with Finchley Road. Take the main path down the center of the cemetery, through the arch beneath the spire of the impressive chapels, and continue straight on until the path reaches a large tree. Turn left, and follow the path for about forty yards until it makes a sharp left turn; at this point continue straight ahead, onto the grass, towards an area that has been overgrown by bushes (below). The grave, number D2–96, is marked by a rough-hewn Celtic cross (a popular design in the late Victorian era) with four trefoil knots, a feature of Celtic art (right). Abbott's wife Mary died in 1919; they had one son, Edwin, and one daughter, Mary, both of whom died in 1952. The gravestone commemorates all four.

While Abbott was alive, he surely believed that his most significant contributions to human culture would be his more serious writings. Posterity, fickle

ABBOTT'S GRAVE IN HAMPSTEAD CEMETERY.

as ever, has decided otherwise. The broad issues that Abbott tackled in his literary and theological works remain important, but the details upon which he expended so much scholarly effort have become as obsolete as the Victorian society in which he lived. *Flatland* has outlasted his other books because it is timeless. A twenty-first-century reader can identify with poor A. Square, and with his lonely battle against mindless orthodoxy and social hypocrisy, as easily as a nineteenth-century one. Abbott's true memorial is not a stone cross in an untidy corner of an obscure cemetery. It is *Flatland* — and unlike a decaying gravestone, its significance increases with every year that passes.

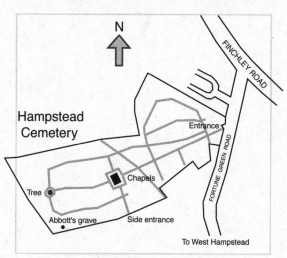

LOCATION OF ABBOTT'S GRAVE.

THE ANNOTATED FLATLAND

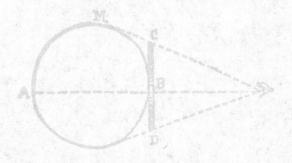

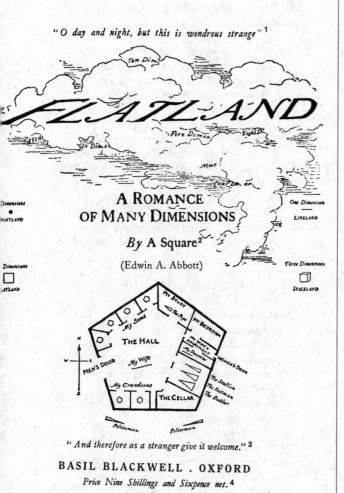

"O day and night, but this is wondrous strange" [1]

FLATLAND

A ROMANCE OF MANY DIMENSIONS

By A Square [2]

(Edwin A. Abbott)

" And therefore as a stranger give it welcome." [3]

BASIL BLACKWELL . OXFORD

Price Nine Shillings and Sixpence net. [4]

1 Spoken by Horatio, in William Shakespeare's *Hamlet,* act 1, scene 5, line 164. Hamlet has just been conversing with his father's ghost, who is now speaking from under the stage.

This is the first of nine quotations from Shakespeare that occur in *Flatland*; no other author is quoted. Abbott was a Shakespeare scholar, who in 1871 published both *Shakespearian Grammar* and *English Lessons for English People* (with Sir John Seeley). On first taking up his post as headmaster at the City of London School in 1865, he wrote, "… I read with my older pupils Shakespeare, *In Memoriam,* Milton, Spenser, Bacon, once also a translation of Dante, and Pope regularly…." (Unless otherwise stated, all quotations about Abbott come from A. E. Douglas-Smith's *City of London School.*) His former pupil, L. R. Farnell, who wrote the article on Abbott in the *Dictionary of National Biography,* recalled that Abbott

> was passionately devoted to English Literature and was himself a considerable Shakespearean scholar; therefore he made the study of a good English book part of the terminal work of all the forms, and hereby I was early brought into contact with many of our masterpieces, especially Shakespeare.

The archives of the City of London School contain an autographed (and bookplated) copy of H. H. Asquith's *Globe Edition of the Works of William Shakespeare* (edited by William George Clark and William Aldis Wright, published by Macmillan in 1867), which Asquith has dated 12 September 1868. Asquith, who later became prime minister, attended the school from 1864 to 1870, so presumably this is the text from which he learned his Shakespeare — and, by inference, from which Abbott taught it.

2 Abbott originally wrote *Flatland* under this pseudonym (first edition of 1884 and second, revised edition, also of 1884). The 1926 reprint by Blackwell adds the author's name. It is hardly necessary to point out that the pseudonym is a pun: the protagonist is "a square." In *The Fourth Dimension* Rudy Rucker suggests that because his middle name was the same as his surname, Edwin Abbott Abbott might have been nicknamed Abbott Squared. If so, he "might have felt a considerable degree of identification with … the hero of *Flatland*." I know of no evidence to support this otherwise attractive conjecture.

Flatland draws its inspiration from school geometry texts, which at that time were universally derived (with few major changes except the omission of difficult or obsolete material) from a classic text: the *Elements* of the ancient Greek geometer Euclid. Squares are defined in Definition 22 of the *Elements*: "Of quadrilateral figures, a square is that which is both equilateral and right-angled." They make their first significant appearance in Proposition 46: *On a given straight line to describe a square.* As a typical illustration of Abbott's source material for the creatures of *Flatland,* I shall reproduce Euclid's solution, following the translation of Sir Thomas L. Heath {my annotations are in braces}. The letters refer to Figure 1. The stilted style and formal logical presentation became synonymous with geometry to generations of schoolboys — a pity because geometric reasoning can be very visual and free-flowing,

which makes it intuitively appealing. "Math anxiety" may well have started here (though algebra, with its *x* and *y* but *no actual numbers,* probably did a lot of damage too).

Let AB be the given straight line; thus it is required to describe a square on the straight line AB.
Let AC be drawn at right angles to the straight line AB from the point A on it [1,11] {meaning *Elements*, Book 1, Proposition 11, where this construction has already been described and proved correct}, and let AD be made equal to AB {which can be done using compasses with the point set at A and the pencil at B: now swing round to D};
Through the point D let DE be drawn parallel to AB {Euclid's Axiom 5 says that this is always possible, and [1,31] explains how to do it},
and through the point B let BE be drawn parallel to AD [1,31].
Therefore ADEB is a parallelogram;
    therefore AB is equal to DE, and AD to BE. [1,34].
But AB is equal to AD;
Therefore the four straight lines BA, AD, DE, EB are equal to one another;
    therefore the parallelogram ADEB is equilateral.
I say next that it is also right-angled.
For, since the straight line AD falls upon the parallels AB, DE,
    the angles BAD, ADE are equal to two right angles. [1,29].
But the angle BAD is right;
    therefore the angle ADE is also right.
And in parallelogrammic areas the opposite sides and angles are equal to one another; [1,34].
    therefore each of the opposite angles ABE, BED is also right.
Therefore ADEB is right-angled.
And it was also proved equilateral.
Therefore it is a square; and it is described on the straight line AB.
Q.E.F.

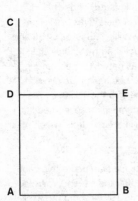

Figure 1 Euclid's construction for a square.

Q.E.F, in Greek, is οπερ εδει ποιησαι (oper edei poiesai); in Latin translation it is Quod Erat Faciendum, "which was to be done." This incantation goes at the end of *porisms,* or "things to be sought." The more familiar Q.E.D. (οπερ εδει δειχαι, oper edei deixai) stands for Quod Erat Demonstrandum, which means "which was to be demonstrated (or proved)," and occurs at the end of *theorems* (where there *is* something to be proved).

3 Hamlet's reply to Horatio in *Hamlet,* act 1, scene 5, line 165. The next two lines are the famous "There are more things in heaven and earth, Horatio, than are dreamt of in your philosophy." Both Shakespeare and Abbott employ the same (rather feeble) word play, strange/stranger, for the same purpose: to encourage the acceptance of something that initially appears bizarre. Hospitable people welcome the stranger, so surely they should welcome the strange.

4 The cover price of the first and second editions is "Half-a-crown." This is 2 shillings and 6 pence (2s.6d.) in old-style British money (20 shillings = 1 pound, 12 pence = 1 shilling, 4 farthings = 2 halfpennies = 1 penny). A crown was 5 shillings. In today's decimal coinage, half a crown is $12\frac{1}{2}$ pence, equivalent to about 18 cents US — but that's not allowing for inflation. Between 1884 and 1984, the purchasing power of the pound divided by about 40, and the last seventeen years have pretty much halved it again. Thus in today's money, the book would have cost about £10 ($15). This suggests that the real price of books has stayed pretty much the same since Abbott's day.

FLATLAND

A Romance of Many Dimensions

With Illustrations
by the Author, A SQUARE
(EDWIN A. ABBOTT)

WITH INTRODUCTION BY
WILLIAM GARNETT, M.A., D.C.L.[5]

"Fie, fie, how franticly I square my talk!"[6]

BASIL BLACKWELL — OXFORD
1926

5 William Garnett (?–1932) was a pupil at the City of London School from 1864 until 1869, and he was present on the day of Abbott's arrival as headmaster in 1865. He was "captain" (top of the school) in mathematics in the same year that H. H. Asquith was captain in classics. Garnett went to Trinity College, Cambridge, where he became chief assistant to the great Scottish physicist James Clerk Maxwell. In the same year that *Flatland* was published, he coauthored with Lewis Campbell a biography of Maxwell (*The Life of James Clerk Maxwell,* Macmillan 1884). The book refers to Maxwell's interest in four-dimensional space:

> My soul is an entangled knot
> Upon a liquid vortex wrought
> The secret of its Untying
> In four-dimensional space is lying.

(Knots can be untied in four dimensions; see *Flatterland,* chapter 4, "A Hundred and One Dimensions.") Garnett became principal of Durham College of Science and educational adviser to the London County Council. His brother Edward, an author and critic, was instrumental in promoting the works of Joseph Conrad (1857–1924) and D(avid) H(erbert) Lawrence (1885–1930). After leaving the school, Garnett maintained contact with Abbott, and upon retirement he moved to Hampstead, where Abbott lived. He signed Abbott's eightieth-birthday message and attended his funeral in 1926. He was therefore the obvious choice to write a preface for the 1926 Blackwell reissue of *Flatland.*

6 Said by Titus in Shakespeare's *Titus Andronicus,* act 3, scene 2, line 31, in a passage concerning the loss of his and Lavinia's hands. The word square had innumerable meanings in Shakespeare's day, including "to regulate," "to make appropriate," "to adapt," "to differ," and (the exact opposite) "to agree." From the context, it seems that Titus is embarrassed because he has unconsciously been adapting his words to reflect the circumstances and has just

realized that this is completely unnecessary because nobody is going to forget that they've just
lost their hands. The quotation is not especially appropriate, and its main relevance to *Flatland* is that
the narrator is A. Square: The Square talks, rather
than the talk being squared.

To

The Inhabitants of SPACE IN GENERAL
and H.C. IN PARTICULAR[7]
This Work is Dedicated
By a Humble Native of Flatland
In the Hope that
Even as he was Initiated into the Mysteries
Of THREE Dimensions
Having been previously conversant
With ONLY TWO
So the Citizens of that Celestial Region
May aspire yet higher and higher
To the Secrets of FOUR FIVE OR EVEN SIX Dimensions
Thereby contributing
To the Enlargement of THE IMAGINATION
And the possible Development
Of that most rare and excellent Gift of MODESTY[8]
Among the Superior Races
Of SOLID HUMANITY

PRINTED IN GREAT BRITAIN IN THE CITY OF OXFORD AT THE ALDEN PRESS
AND BOUND BY THE KEMP HALL BINDERY, OXFORD

7 H.C. is Howard Candler (?–1916), Abbott's best and lifelong friend. Candler was mathematics master at Uppingham School, where for some years Charles Hinton was science master. In the introduction to his theological book *The Fourfold Gospel,* written just after Candler died, Abbott explicitly states that Candler was "H. C." The first edition of Flatland in the library of Trinity College, Dublin, donated by one of Candler's grandsons, is inscribed with the handwritten message "To H.C., in particular."

8 In *The Spirit on the Waters* (1897), Abbott relates the climax of *Flatland,* in which A. Square experiences a visitation from the third dimension — the Sphere, perceived as a series of ever-changing Circles. Would A. Square be right in worshipping the Sphere because of its God-like powers? No, Abbott tells his readers. It is wrong to attribute spiritual or moral superiority to a being merely because of its physical or mental abilities. And he enlarges on this section of *Flatland*'s Dedication:

> This illustration from four dimensions … may serve a double purpose in our present investigation. On the one hand it may lead us to vaster views of possible circumstances and existence; on the other hand it may teach us that the conception of such possibilities cannot, by any direct path, bring us closer to God. Mathematics may help us to measure and weigh the planets, to discover the materials of which they are composed, to extract light and warmth from the motion of water and to dominate the material universe; but even if by these means we could mount up to Mars, or hold converse with the inhabitants of Jupiter or Saturn, we should be no nearer to the divine throne, except so far as these new experiences might develop our modesty, respect for facts, a deeper reverence for order and harmony, and a mind more open to new observations and to fresh inferences from old truths.

Introduction

In AN ADDRESS to the Committee of the Cayley[9] Portrait Fund in 1874[10] Clerk Maxwell,[11] after referring in humorous terms to the work of Arthur Cayley in higher algebra and algebraical geometry, concluded his eulogium[12] with the lines—

March on, symbolic host! with step sublime,
Up to the flaming bounds of Space and Time!
There pause, until by Dickenson depicted,[13]
In two dimensions, we the form may trace
Of him whose soul, too large for vulgar space
In n *dimensions flourished unrestricted.[14]*

In those days any conception of "dimensions" beyond length breadth and height was confined to advanced mathematicians; and even among them, with very few exceptions, the fourth and higher dimensions afforded only a field for the practice of algebraical analysis with four or more variables instead of the three which sufficiently describe the space to which our footrules are applicable. Any geometrical conclusions reached were regarded only as analogies to the corresponding results in geometry of three dimensions and not as having any bearing on the system of Nature. As an illus-

9 Arthur Cayley (1821–1895), English mathematician, educated at Cambridge University, becoming a fellow of Trinity College in 1845 but leaving after three years because he did not wish to take holy orders. He then spent fifteen years as a lawyer before being appointed the first Sadleirian Professor of Mathematics at Cambridge in 1863. Apart from half a year spent at Johns Hopkins University in the United States (1881–1882), he remained at Cambridge until his death. Cayley is best known for his work with James Joseph Sylvester (1814–1897) on *invariants,* algebraic expressions that remain unchanged when their variables are transformed. The theory of invariants, generalized to differentials of the variables, formed the mathematical basis of Albert Einstein's General Theory of Relativity — the idea that gravity is a manifestation of the curvature of a four-dimensional space-time. Cayley was also responsible for *matrix algebra,* which is now widely used in all branches of pure and applied mathematics: A matrix is a rectangular table of numbers and represents a transformation of variables. Most significantly for this book, Cayley was one of the creators of higher-dimensional geometry, beginning with a paper of 1845 on spaces of *n* dimensions. The main founder of this theory was the German mathematician Hermann Grassman (1809–1877), in his *Ausdehnungslehre* (*Theory of Extension*) of 1844, but Cayley's early work was done independently.

10 The portrait concerned, which was painted by Lowes Cato Dickinson (1819–1908) in 1874, is in the possession of Trinity College, Cambridge. It is shown in the frontispiece to volume VI of *The Collected Mathematical Papers of Arthur Cayley Sc.D., F.R.S.,* Cambridge University Press 1893 (13 volumes and an index). The "portrait fund" was set up by a group of private individuals and was not an official activity of Trinity College; however, the group presented the portrait to the college and probably included several fellows of Trinity. The membership of the group seems not to be known, but the college archives state that the portrait was "given by the subscribers through Mr Walton Chairman of the Cayley Portrait Committee, Apr, 1874." It measures 43 inches by 33½ inches, and is described as follows:

> Life size seated figure — to knee. Head three quarters to rt [right], features [are] elderly. Long wavy brown hair at sides — head nearly bald on crown. Figure three quarters to rt, seated at a sloping desk, right hand holding quill pen to sheet of paper lying on the desk. Left hand resting on right wrist. Inkstand on desk. Drawer at end of desk partly open showing papers & sealing wax. Wears black coat, M.A. [Master of Arts] gown, soft white collar & dark necktie.
>
> On canvas: LCD [monogram of Lowes Cato Dickinson] 1873.
>
> On frame: Arthur Cayley Sc.D. F.R.S. Painted by Lowes Dickinson 1874. Presented to Trinity College by the Subscribers.

F.R.S. stands for Fellow of the Royal Society, Britain's most prestigious scientific organization (then and now). The disagreement in dates suggests that the painting was done in 1873 but not presented until 1874.

11 James Clerk Maxwell (1831–1879), Scottish mathematical physicist who revolutionized technology with Maxwell's Equations for electricity and magnetism, which he devised in 1864 to provide a mathematical basis for the discoveries of Michael Faraday (1791–1867). The existence of electromagnetic waves (and hence radio, television, and radar) is a direct consequence of Maxwell's Equations, and such waves were discovered as a result of this mathematical prediction. Maxwell published his first scientific paper at the age of 14 and obtained a degree in mathematics from Trinity College, Cambridge, in 1854. In 1856 he became professor of natural philosophy at Marischal College, Aberdeen; in 1860 he moved to King's College,

tration, reference may be made to the "more divine offspring of the divine Cube in the Land of Four Dimensions" mentioned on p. 171 *infra* which has for its faces eight three-dimensional cubes and possesses sixteen four-dimensional angular points or corners.

During the present century the work of Einstein,[15] Lorentz,[16] Larmor,[17] Whitehead[18] and others has shewn that at least four dimensions of *space-time*[19] are necessary to account for the observed phenomena of nature, and there are some suggestions of the necessity for more than four. It is only when dealing with very high velocities,[20] such as are comparable with the velocity of light, that the unity of time with space thrusts itself upon the notice of physicists, for even with such a velocity as that of the planet Mercury in its orbit[21] it is only after the lapse of centuries that any divergence from the motion strictly calculated on the basis of Euclidean Geometry[22] and Newton's[23] laws of gravitation and of motion has become apparent. The observed behaviour of electrons, moving in high vacua with velocities comparable with the velocity of light, has confirmed some of Einstein's conclusions and necessitated a revision of our fundamental notions of kinematics and the laws of motion when these high velocities are concerned. But the whole subject of Relativity has strongly appealed to popular interest through the brilliant confirmation of Einstein's theory of gravitation by the bending of light[24] in passing close to the sun's surface and the consequent apparent displacement of stars which are very close to the sun from their true relative position when photographed during a solar eclipse. The best popular exposition of the whole subject of relativity and gravitation is to be found in Professor Eddington's[25] *Space, Time, and Gravitation.*

But when a great truth comes to light it is generally found that there have already been prophets crying in the wilderness and preparing the way for the reception of the Revelation when the full time has come. In an anonymous letter published in *Nature*[26] on February 12th, 1920, entitled "Euclid, Newton, and Einstein," attention was called to such a prophet in the following words:—

"Some thirty or more years ago, a little *jeu d'esprit* was written by Dr. Edwin Abbott, entitled 'Flatland.' At the time of its publication it did not attract as much attention as it deserved. Dr. Abbott pictures intelligent beings whose whole experience is confined to a plane, or other space of two dimensions, who have no faculties by which they can become conscious of anything outside that space and no means of moving off the surface on which they live. He then asks the reader, who has the consciousness of the third dimension, to imagine a sphere descending upon the plane of Flatland and passing through it. How will the inhabitants regard this phenomenon? They will not see the approaching sphere and will have no conception of its solidity. They will only be conscious

London; in 1871 he became the first Cavendish Professor of Physics at Cambridge University. Maxwell's interests were broad, as was typical of the "Scottish Enlightenment" period. He worked on color vision (his theories influenced several Scottish painters, and he produced one of the earliest color photographs), Saturn's rings, mechanics, and the kinetic theory of gases (which explains such phenomena as temperature and pressure in terms of the chaotic motion of gas molecules). His main relevance to Garnett's introduction to *Flatland* is his contribution to the development of multidimensional geometry.

12 The arts and sciences in the nineteenth century were more closely related than they are today, and it was not uncommon for a scientific paper or address to include verse. One of the great exponents of such scientific poetry was Cayley's collaborator Sylvester, who wrote a pamphlet called *The Laws of Verse* in 1870. Sylvester provides an important link between *Flatland* and H. G. Wells's famous story *The Time Machine* of 1895 (discussed later), and he was generally an oddball character. He applied for the Gresham Professorship of Geometry (see the preface to this book) but failed to secure the post. After some difficulty he obtained the position of professor of mathematics at the Royal Military Academy, Woolwich, until he was forcibly retired in 1870 at the age of fifty-six because he was superannuated —the official term for someone who was too old to be of further use. Such retirement might have made sense for soldiers, but it made none for Sylvester, who was approaching the pinnacle of his intellectual powers. In 1876 he moved to the United States to take a founding professorship at the new Johns Hopkins University. There he remained until 1883, when he was persuaded to accept the newly vacant Savilian Chair of Geometry at Oxford University. During his inaugural lecture, he broke into verse in order to emphasize the strange absence of a particular term in an algebraic expression:

of the circle in which it cuts their plane. This circle, at first a point, will gradually increase in diameter, driving the inhabitants of Flatland outwards from its circumference, and this will go on until half the sphere has passed through the plane, when the circle will gradually contract to a point and then vanish, leaving the Flatlanders in undisturbed possession of their country.... Their experience will be that of a circular obstacle gradually expanding or growing, and then contracting, and they will attribute to *growth in time* what the external observer in three dimensions assigns to motion in the third dimension. Transfer this analogy to a movement of the fourth dimension through three-dimensional space. Assume the past and future of the universe to be all depicted in four-dimensional space and visible to any being who has consciousness of the fourth dimension. If there is motion of our three-dimensional space relative to the fourth dimension, all the changes we experience and assign to the flow of time will be due simply to this movement, the whole of the future as well as the past always existing in the fourth dimension."

It will be noticed that in the presentation of the Sphere to the Flatlander the third dimension involves time through the motion of the Sphere. In the Space-Time Continuum of the Theory of Relativity the fourth dimension is a time function, and the simplest element is an "event." One set of parallel sections of the four-dimensional contin-

TO A MISSING MEMBER
OF A FAMILY OF TERMS IN AN ALGEBRAICAL FORMULA

Lone and discarded one! divorced by fate,
From thy wished-for fellows—whither art flown?
Where lingerest thou in thy bereaved estate,
Like some lost star or buried meteor stone?
Thou mindst me much of the presumptuous one
Who loth, ought less than greatest, to be great
from Heaven's immensity fell headlong down
to live forlorn, self-centred, desolate:
Or who, new Heraklid, hard exile bore,
Now buoyed by hope, now stretched on rack of fear,
Till throned Astraea, wafting to his ear
Words of dim portent through the Atlantic roar,
Bade him "the sanctuary of the Muse revere
And strew with flame the dust of Isis' shore."

He then resumed his mathematical discussion.

13 The spelling is wrong: The portrait of Cayley was painted by Lowes Cato Dickinson. Dickinson painted many Victorian intellectuals and celebrities, among them Charles Kingsley (1819–1875, portrait 1862), a clergyman who accepted Darwin's theory of evolution and was thereby inspired to write *The Water-Babies* in 1863; the novelist Mary Ann Evans (George Eliot, portrait 1872); Sir Charles Lyell (1797–1875, portrait — a replica — 1883), the geologist who discovered "deep time"; and the whole of Gladstone's cabinet (painted 1869–1874).

14 The *logical* basis of the mathematics of *n*-dimensional space is straightforward and does not depend on the properties of actual physical space. The *psychological* ramifications are more convoluted. Cayley formulated the main ideas clearly in 1845 (recall that Grassmann had done so independently the year before), although their prehistory goes back much further. The geometry of *n* dimensions is defined by analogy with two and three dimensions but then takes on a life of its own. In Cartesian coordinate geometry, the plane is represented as the set

uum present the universe as it exists in three-dimensional space at the instants corresponding to the sections. Sections in all other directions involve the time element and represent the universe as it appears to an observer in motion.

There are some mathematical minds which are completely satisfied by the results expressed in algebraical symbols of the analysis of a continuum of four dimensions; but there are others which crave for the visualization of these results[27] which, in their symbolic forms, they do not question. To many, perhaps to the great majority, of these, Dr. Abbott's sphere penetrating Flatland points the way to the clearest imagery of the fourth dimension to which they are likely to attain.

WM. GARNETT.

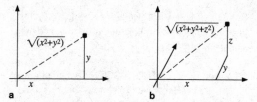

Figure 2 (a) Distances in two-dimensional space. **(b)** Distances in three-dimensional space.

of all points with coordinates (x, y), where x and y are positive or negative numbers. The distance between two points (x, y) and (u, v) can be deduced from the Pythagorean theorem and is given by the formula

$$\sqrt{(x-u)^2 + (y-v)^2}$$

See Figure 2a. Similarly, three-dimensional space is represented as the set of all points with coordinates (x, y, z), and the distance between two points (x, y, z) and (u, v, w) is given by the formula

$$\sqrt{(x-u)^2 + (y-v)^2 + (z-w)^2}$$

See Figure 2b. The algebraic generalization to n dimensions is natural and, to modern tastes, inevitable: A point in n-dimensional "space" is considered to be represented by an n-tuple of numbers $(x_1, ..., x_n)$, and the distance between that point and another one $(w_1, ..., w_n)$ is defined to be

$$\sqrt{(x_1-w_1)^2 + ... + (x_n-w_n)^2}$$

All the main features of Euclidean geometry can be expressed purely in terms of distances, so these features can be extended to n-dimensional space by applying the above formula. The generalization is purely formal, so the "space" that it defines has no specific *physical* interpretation — but its mathematics can readily be worked out. Only later did it become clear that this kind of multidimensional "geometry" is widely applicable — for example in mechanics, statistics, and economics. By the 1960s, mathematicians had become so used to the concept that working in n-dimensional space was a natural reflex.

See "The Fourth Dimension in Mathematics" in this book for a history of the mathematics of multidimensional spaces.

15 Albert Einstein (1879–1955), German physicist famous for his Special and General Theories of Relativity, which overthrew the physical theories of Sir Isaac Newton. Einstein was born in Ulm and educated in Zürich. In one year, 1905, he published four groundbreaking papers in widely differing areas of physics. In 1919 his prediction that a gravitational field can bend light was verified by observations of stars made during a solar eclipse by a team headed by Sir Arthur Stanley Eddington. Einstein was awarded the Nobel Prize in 1921. Fleeing Nazi Germany, he moved to the Institute for Advanced Study in Princeton in 1933 and became a U.S. citizen in 1940.

According to the Special Theory of Relativity, no material particle can travel faster than light (186,000 miles, or 300,000 kilometers, per second). Special relativity includes Einstein's celebrated formula $E = mc^2$, which expresses the conversion of matter m into energy E. Here c is the speed of light, which (as we have just seen) is very large, and its square c^2 is even larger, implying that a small amount of matter "contains" a huge amount of energy. This equation is fundamental to nuclear power and the atomic bomb, although Einstein himself was a pacifist. According to the General Theory of Relativity, gravity is a result of the curvature of space-time. One spectacular consequence of general relativity is the existence of black holes — regions of space-time from which nothing, not even light, can escape. It is now believed that most galaxies have giant "supermassive" black holes at their cores.

The mathematics of relativity, both special and general, rests on considering space and time to form a single, four-dimensional "space-time continuum." Until the mid-twentieth century, this was the principal — and without doubt the best-known —

application of multidimensional geometry. See *Flatterland* chapter 12 ("The Paradox Twins") and the first part of chapter 13 ("Domain of the Hawk King") for a treatment of special relativity. See the rest of chapter 13 and chapter 14 ("Down the Wormhole") for general relativity.

16 Hendrik Antoon Lorentz (1853–1928), Dutch physicist who won the Nobel Prize in 1902 for his theory of electromagnetism, which gave rise to Einstein's Special Theory of Relativity. Lorentz was born in Arnhem and became professor of mathematical physics at Leiden in 1878. The central aim of his work was to devise a unified theory of electricity, magnetism, and light, improving on that of Maxwell. He predicted that a strong magnetic field should affect the wavelength of light emitted from atoms, and in 1896 his pupil Pieter Zeeman (1865–1943) verified the prediction experimentally, in what is now called the Zeeman effect. The Nobel Prize was awarded to Lorentz and Zeeman for this work. However, Lorentz's theory failed to explain the experiments of the American physicist Albert Abraham Michelson (1852–1931) and the American chemist Edward Williams Morley (1838–1923), who in 1887, while testing the theory that electromagnetic radiation was conveyed by the "luminiferous ether," showed that the motion of the Earth relative to the alleged ether was negligible. Lorentz was led to suggest that local time depends on the velocity of the observer. In conjunction with a similar suggestion by the Irish physicist George Francis FitzGerald (1851–1901), in which the length of a body is held to contract as its speed increases, Lorentz formulated his "Lorentz transformations," upon which Einstein based the Special Theory of Relativity. See *Flatterland* chapter 12.

17 Sir Joseph Larmor (1857–1942), Irish physicist born in Magheragall, County Antrim. Larmor calculated the rate at which an accelerated electron radiates energy and offered the first successful explanation for the splitting of spectral lines by a magnetic field. His work was based on the (now discredited) idea that matter is composed entirely of electrical particles moving through the ether. He taught at Queen's College, Galway, from 1880 to 1885 and at Cambridge from 1885 to 1932. He was knighted in 1909 and served as a member of parliament (for Cambridge University) from 1911 to 1922.

18 Alfred North Whitehead (1861–1947), English mathematical logician and philosopher. Born in Ramsgate, Whitehead is best known for his collaboration with the English mathematician and philosopher Bertrand Russell (1872–1970) on *Principia Mathematica,* an impressively technical three-volume work on the logical foundations of mathematics that was published between 1910 and 1913. Whitehead became a fellow of Trinity College, Cambridge, in 1884, moved to London in 1910, and secured a position at University College London in 1911. In 1914 he was made professor of applied mathematics at Imperial College of Science and Technology, London. In 1924 he moved to the United States, becoming professor of philosophy at Harvard University. His relevance to Garnett's introduction to *Flatland* stems from his interest in the philosophy of science: For the first half of the 1920s, he devoted much effort to constructing philosophical foundations for physics, seeking to understand the perceptual aspects of relativity. His books *Enquiry Concerning the Principles of Natural Knowledge* and the more popular *The Concept of Nature* date from this period. In 1925 he emphasized the distinction between a mathematical description of nature in terms of matter, energy, and motion, and the actual concrete reality *described* by the mathematics. These ideas were published as *Science and the Modern World,* and he returned to the theme in his *Adventures of Ideas* in 1933.

19 The geometrical basis of relativity is the idea that the traditional three dimensions of space should be augmented by a fourth: time. In Newtonian physics, space is considered to be the standard three-dimensional space of Euclid, in which the position of a point can be defined by three coordinates (x, y, z) relative to three specified axes (east-west, north-south, up-down). Time t can be thought of as a fourth coordinate, leading to a four-dimensional structure with coordinates (x, y, z, t). This idea goes back at least to Jean le Rond D'Alembert (1717–1783), who suggested thinking of time as a fourth dimension in his article on "dimension" in the *Encyclopédie ou Dictionnaire Raisonné des Sciences, Arts, et des Métiers* (*Reasoned Encyclopedia or Dictionary of Sciences, Arts, and Crafts*), published 1751–1780. Joseph-Louis Lagrange (1736–1813) used time as a fourth dimension in his *Mécanique Analytique* (*Analytical Mechanics*) of 1788 and again in his *Théorie des Fonctions Analytiques* (*Theory of Analytic Functions*) of 1797, in which he says, "Thus we may regard mechanics as a geometry of four dimensions." However, Einstein's innovation goes deeper. In Newtonian physics, the time coordinate is physically very different from the space coordinates. In relativity, spatial and temporal coordinates can be transformed into each other when the laws of physics are expressed relative to a moving coordinate system — a moving *observer.* The basic geometry of special relativity was introduced by the Russian-born German mathematician Hermann Minkowski (1864–1909) in a paper of 1908, wherein the terms *world line* and *light cone* first appear (see *Flatterland* chapter 13). Minkowski reformulated Maxwell's Equations in tensor form, a step that Einstein at the time denounced as "superfluous learnedness." By 1912 Einstein had changed his mind and was employing tensor methods himself; in 1916 he stated that Minkowski's first steps in that direction had greatly eased the transition from special to general relativity.

In *Stella,* and more extensively in *An Unfinished Communication,* C. H. Hinton develops the remarkable idea of two-dimensional time. Represent a being's life as a fixed world line in conventional four-dimensional space-time. But now imagine that the world line can vary along a *fifth* dimension. In such a setting, the future of an event is not uniquely determined, so this idea can be seen as an early anticipation of the "parallel worlds" or "alternat(iv)e universes" trope of science fiction, which was introduced into mainstream physics by Hugh Everett in his "many worlds" interpretation of quantum reality. See John Gribbin, *In Search of Schrödinger's Cat.*

20 The difference between Newtonian physics and relativity becomes apparent only for bodies (or observers) that are moving with a velocity that is an appreciable percentage of the speed of light. This is why we do not notice relativistic effects in everyday life. It is also why Newtonian physics was good enough for nearly all scientific purposes until the twentieth century — and is still good enough for most of them in the twenty-first.

21 The first observational evidence for the General Theory of Relativity antedated that theory by many years. According to Johannes Kepler (1571–1630) and Isaac Newton, planetary orbits are (to an excellent approximation) ellipses. This and two other "laws" of planetary motion discovered by Kepler led Newton to formulate his Inverse-Square Law of Gravitation: Every particle in the universe attracts every other particle with a force that is proportional to their masses and inversely proportional to the square of the distance between them. This law is still used for most astronomical purposes. Observations show that the *perihelion* of Mercury's orbit — its point of closest approach to the Sun — *precesses.* That is, its position in space (relative to the sun) slowly rotates. In the framework of Newtonian physics, calculations show that most of the precession is caused by perturbations from other planets in the solar system (especially Jupiter and Saturn, the most massive ones), but even when these perturbations are taken into account, a discrepancy re-

mains. The first person to notice this and to offer an explanation was the French astronomer Urbain Jean Joseph Le Verrier (1811–1877), who in 1859 published the text of a letter to Hervé Faye (1814–1902) in which he stated that the perihelion of Mercury advances by 38 seconds of arc per century. How could this be explained? It would make sense if the mass of Venus were increased by at least 10 percent, but that had to be ruled out on other grounds. A new planet inside Mercury's orbit could also do the trick, but that was unlikely. Le Verrier postulated a swarm of asteroids inside Mercury's orbit.

In 1882 the American astronomer Simon Newcomb (1835–1909) improved Le Verrier's measurement and found an advance of 43 seconds of arc per century, a figure that has not changed significantly since. This advance of Mercury's perihelion is exceedingly tiny, but celestial mechanics is a very precise area. Possible explanations included a planetary ring inside Mercury's orbit and an unnoticed moon of Mercury. Newcomb himself came "to prefer provisionally the hypothesis that the Sun's gravitation is not exactly as the inverse square." In 1916 Einstein calculated the rate of precession on the basis of the General Theory of Relativity — in which Newcomb's hypothesis is correct — and exactly accounted for the discrepancy. Because Mercury is very close to the sun, it experiences a much stronger gravitational field than any other planet in the solar system; the amount of perihelion precession for the remaining planets is so small that it is swamped by other variations in the orbit, which are caused by perturbations by other planets.

In 1967 the American physicist Robert H(enry) Dicke (1916–) suggested that part of the 43 seconds could be explained in Newtonian physics by the flattening of the sun's poles, opening up the possibility that relativity might need to be modified because Einstein's calculations now explained *too much.* However, Einstein's original version was supported by work of the American astronomer Ronald W. Hellings on the motion of the solar system and by that of the American astronomer Joseph H. Taylor on the binary pulsar system PSR 1913+16.

22 Eucleides, anglicized as Euclid (365?–275? B.C.), Greek geometer, author of the *Elements,* a treatise on geometry in thirteen volumes. Euclid deduced a large part of Greek geometry (of the plane and space) from a small number of definitions ("point," "line," and so on) and five *axioms* — statements whose validity would be assumed at the outset, such as "all right angles are equal" and "any two points can be joined by a line."

Later generations of mathematicians found flaws in Euclid's presentation (additional unstated assumptions, such as "if a line passes through one side of a triangle, then it will meet another side if sufficiently extended"). In his *Grundlagen der Geometrie* (*Foundations of Geometry*) of 1899, the German mathematician David Hilbert (1862–1943) wrote down a system of axioms (this system was repeatedly revised — there were twenty axioms in the seventh edition of 1930) and derived Euclid's geometry from them in full logical rigor. Nonetheless, Euclid did an amazingly good job for 300 B.C. In particular, he saw the need for his infamous Parallel Axiom (axiom 5): "If a straight line falling on two straight lines makes the interior angles on the same side less than two right angles, the two straight lines, if produced indefinitely, meet on that side where the angles are less than two right angles." (See Figure 3.) Euclid's Parallel Axiom is logically equivalent to the following one, which is simpler-looking, though still complicated: "Given a line, and any point not on

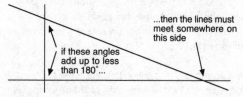

Figure 3 Euclid's Parallel Axiom.

that line, there exists one and only one line parallel to the given line and passing through the given point." This reformulation is generally known as *Playfair's Axiom,* after the English mathematician John Playfair (1748–1819), who stated it in 1795, but in fact it was known to Proclus (410–485), a Greek commentator on Euclid. For most of the next 2,000 years, many people thought that the Parallel Axiom was superfluous and tried to deduce it from the other four. Only around 1830 did the invention of non-Euclidean geometry prove that no such deduction is possible and that Euclid had done the right thing after all. Non-Euclidean geometry has a long and complex history and prehistory, but the definitive breakthrough stemmed from the work of three mathematicians:

- The German Carl Friedrich Gauss (1777–1855) between 1792 and 1816
- The Hungarian János Bolyai (1802–1860), who worked his ideas out around 1825 and published them in 1832
- The Russian Nikolai Ivanovich Lobachevskii (1793–1856), who submitted his work to his home university in 1826 and published it in 1829

No original manuscripts by Euclid survive (although his book became one of the most widely copied, and most influential, texts ever), and little is known for certain about his life. We do know that he founded and taught at a school in Alexandria, Egypt, during the reign of Ptolemy I Soter (323–285/283 B.C.). About 800 years later, the Greek philosopher Proclus related one anecdote about Euclid: He is alleged to have told Ptolemy that "there is no royal road to geometry" — no shortcut to save the king effort or time. Supposedly also, when asked what was the point of learning geometry, he told his slave to offer a student money, on the grounds that "he must needs make gains by what he learns." Until the twentieth century, "geometry" in English schools meant a simplified version of the early parts of Euclid's *Elements.* It is this material that provided such fertile ground for Abbott's imagination. Every school-boy — though not every schoolgirl — would be only too familiar with the Euclidean world of a flat plane inhabited by Triangles, Circles, and Squares.

23 Sir Isaac Newton (1642–1727), English mathematician, physicist, alchemist, and — according to the economist John Maynard Keynes — last of the Magi. He was born the same year Galileo died, in the manor house of Woolsthorpe, near the village of Colsterworth, a few miles from Grantham in Lincolnshire. Newton became one of the three greatest mathematicians of all time, the others being Archimedes (287–212 B.C.) and Gauss. His most influential work is the *Philosophiae Naturalis Principia Mathematica* (*Mathematical Principles of Natural Philosophy*) of 1687, revised in 1713 and again in 1726, which set out "the system of the world." Building on the work of predecessors such as Kepler and Galileo Galilei (1564–1642), Newton developed a new approach to nature in terms of underlying mathematical laws based on rates of change — now called *differential equations*. His most important discoveries were his three Laws of Motion and his Law of Gravity. He is also famous for co-inventing calculus, along with the German mathematician and philosopher Gottfried Wilhelm Leibniz (1646–1716), and for the heated controversy over priority that ensued. Newton also discovered many fundamental principles in optics, dabbled in alchemy, tried to date the events in the Bible, and served as master of the Royal Mint.

24 One of the predictions of the General Theory of Relativity is that gravity bends light by twice the amount that Newton's laws imply. In 1919 this prediction was confirmed when Eddington (see next note) led an expedition to Príncipe Island in West Africa, where a total eclipse of the sun was due to occur. A second expedition to Sobral, in Brazil, was led by Andrew Crommelin (1865–1939) of Greenwich Observatory. The expeditions observed stars near the edge of the sun during the period of totality (only during an eclipse would these stars not be swamped by the light of the sun). The observers found slight displacements in the stars' apparent positions — a result consistent with the relativistic predictions. Einstein sent his mother a postcard to relay the news: "Dear Mother, joyous news today. H. A. Lorentz telegraphed that the English expeditions have actually demonstrated the deflection of light from the Sun." The amount of bending predicted by a Newtonian theory of gravity was 0.87", whereas Einstein's prediction was double that: 1.74". (Here the symbol " means "seconds of arc," 1/3600 of a degree.) The Sobral expedition measured 1.98" ± 0.30" and the Príncipe expedition 1.61" ± 0.30". The *Times* of 7 November 1919 ran the following headline: "REVOLUTION IN SCIENCE. NEW THEORY OF THE UNIVERSE. NEWTONIAN IDEAS OVERTHROWN." Halfway down the second column is the subheading "SPACE 'WARPED'." Einstein became a celebrity overnight.

25 Sir Arthur Stanley Eddington (1882–1944), English astronomer, physicist, and mathematician. Born in Kendal, Eddington was both a scientist and a popularizer of science. His greatest work was in astrophysics — the structure and evolution of stars. In 1913 he became Plumian Professor of Astronomy at Cambridge University and in 1914 was appointed Director of its observatory. From 1906 to 1913 he was also chief assistant at the Royal Greenwich Observatory in London. His 1923 book *The Mathematical Theory of Relativity* was considered by Einstein to be the best exposition of the subject in any language. From the late 1920s onward, Eddington also published expository books on science for the general public. *Space, Time, and Gravitation* was published in 1920.

26 Garnett fails to mention that this letter to *Nature*, referring to Abbott as a prophet of space-time, was signed "W. G." By a strange coincidence, these are

Garnett's own initials. Banchoff points out that, if we really need any further confirmation, the relatively unusual phrase *jeu d'esprit* in the letter's opening sentence appears twice in Garnett and Campbell's biography of Maxwell.

27 The technique of *visualization* — representing concepts by intuitively accessible geometric images — is widely used in mathematics. In 1637 René Descartes (1596–1650), in an appendix to his *Discours de la Méthode* (*Discourse on Method*), described the technique of Cartesian coordinates: representing points in the plane by pairs (x, y) of numbers defined by two mutually perpendicular axes (Figure 4a). Similarly, points in three-dimensional space can be represented by triples (x, y, z) defined by three mutually perpendicular axes (Figure 4b). (In fact, Descartes also considered systems of *oblique* axes, which need not meet at right angles.) Cartesian coordinates allow algebraic concepts (in the variables x, y, z) to be interpreted as geometric forms. For example, the equation

$$x^2 + y^2 = 1$$

corresponds to a circle of unit radius, centered at the origin $(0,0)$. The link is the Pythagorean theorem about right triangles: "The square on the hypotenuse of a right triangle is equal to the sum of the squares on the two adjacent sides" (*Elements* [1,47], the Proposition immediately following the construction of a square). This theorem implies that every point (x, y) on the unit circle satisfies the above equation

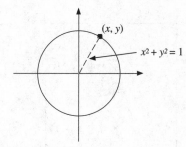

Figure 5 How to turn a circle into an algebraic equation.

and, conversely, any point that satisfies the equation lies on the circle (see Figure 5).

The technique has developed beyond all recognition since then. In Banchoff's words,

> What Abbott and other 19th-century writers envisioned has become a reality in our present day. Encounters with phenomena from the fourth and higher dimensions were the fabric of fantasy and occultism. People (other than the spiritualists) did not expect to see manifestations of four-dimensional forms any more than they expected to encounter Lilliputians or Mad Hatters. Today, however, we do have the opportunity not only to observe phenomena in four and higher dimensions, but we can also interact with them. The medium for such interaction is computer graphics.... Unlike its human operator, a computer has few preconceptions about what dimension it is in. Just as easily as it keeps track of three coordinates for each point, it can, when properly programmed, keep track of four or more coordinates. Often a fourth coordinate can indicate some property of the point on the screen, like color or brightness. At other times it can represent a fourth spatial coordinate, interchangeable with the other three, just as the length, width, and height of a box can be manipulated in three-space.... Thus we use all of our experience with interpreting two-dimensional images of three-dimensional objects to help us move up one further step to interpret the three-dimensional representations of objects which require a fourth coordinate for their effective description.

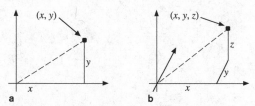

Figure 4 (**a**) Cartesian coordinate system on the plane. (**b**) Coordinate system in three-dimensional space.

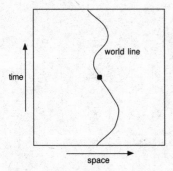

time

world line

space

Figure 6 Minkowski's geometry of relativistic space-time. Here space is schematically represented as being one-dimensional.

In 1908 Minkowksi applied Descartes's idea to physics, introducing a geometric image for *space-time* in which time *t* is explicitly represented as an extra coordinate axis. For example, in cases where only one dimension *x* of space is relevant, space-time can be depicted as a plane (Figure 6) with one spatial axis and one temporal axis. In this interpretation, the path of a moving body in space is visualized as a single, unchanging *path* in space-time. For example, suppose that a point particle moves through space, occupying position $x(t)$ at time *t*. The corresponding set of points $(x(t), t)$ forms a curve in space-time, the *world line* of the particle. This geometric image of space-time is extremely useful in relativity (see "The Fourth Dimension in Mathematics" in this book).

Philosophically, the "world line" image raises an interesting question about free will. In the mathematical formulation of relativity, a particle's world line is a single, unchanging, complete object. Now, a human being is composed of innumerable particles (atoms or subatomic particles), each of which has its own world line. If humans have free will, then their particles' world lines must "unfold" as time passes: At any instant, they exist in the past but not (yet) in the future. Therefore, if the world line has the properties assumed in the mathematical formu-

lation of relativity, humans do not possess free will but only the illusion of free will. However, the interpretational link between mathematical descriptions of reality and reality itself is subtle, and not every concept employed in the mathematical formulation need have a direct real-world meaning, so this argument is open to many objections. The issue is central to the "process philosophy" introduced by Henri (–Louis) Bergson (1859–1941), who wrote to William James that "I saw, to my astonishment, that scientific time does not *endure*." This led him to see reality as a system of ongoing *processes,* not *fixed things.* His 1899 *Time and Free Will* summarizes his views, which were taken up by Whitehead.

Preface to the Second and Revised Edition, 1884.

By the Editor

IF MY POOR Flatland friend retained the vigour of mind which he enjoyed when he began to compose these Memoirs, I should not now need to represent him in this preface, in which he desires, firstly, to return his thanks to his readers and critics in Spaceland, whose appreciation has, with unexpected celerity, required a second edition of his work; secondly, to apologize for certain errors and misprints (for which, however, he is not entirely responsible); and, thirdly, to explain one or two misconceptions. But he is not the Square he once was.[28] Years of imprisonment, and the still heavier burden of general incredulity and mockery, have combined with the natural decay of old age to erase from his mind many of the thoughts and notions, and much also of the terminology, which he acquired during his short stay in Spaceland. He has, therefore, requested me to reply in his behalf to two special objections, one of an intellectual, the other of a moral nature.

The first objection[29] is, that a Flatlander, seeing a Line, sees something that must be *thick* to the eye as well as *long* to the eye (otherwise it would not be visible, if it had not some thickness); and

28 Abbott retired (from the headmastership of the City of London School) in 1889 at age fifty. This is (and was in Victorian times) unusually early. A. Square often speaks with Abbott's voice, and by 1884 Abbott — who was slightly built, though generally healthy — may well have felt that he was "not the man he once was." The school retains a copy of his letter of resignation (Figure 7), from an original held in the Corporation of London Records Office, Common Council papers, 14 March 1889.

> City of London School
> Victoria Embankment E C
> 13 March 1889

To the Right Hon[ble] the Lord Mayor.
My Lord Mayor,

I shall be obliged by you submitting to the Court of Common Council my resignation of the Headmastership of the City of London School to take effect next Michaelmas.

The responsibilities of the office must, under any circumstances, heavily tax the energy as well as the intellectual qualifications of the Headmaster for the time being: and the burden has been increased by the important changes in the course of study introduced by the Court last year, changes which — though beneficial in themselves, if well supervised controlled and directed — require increased rather than failing powers in the Headmaster, to supervise, and possibly to modify or develop them.

To discharge these responsibilities, after upwards of three and twenty years of service, I no longer feel fully competent. Were I to attempt the task, the kindness of the Court towards one who has served them for a considerable period to the best of his ability, would, I dare say, tolerate me for another five, or even for another ten years. But the past history and the high reputation of the school preclude me from thus trespassing upon tolerance. One who is content to be tolerated is probably not fit to be the Headmaster of any school, and certainly not of that which bears the name of the City of London.

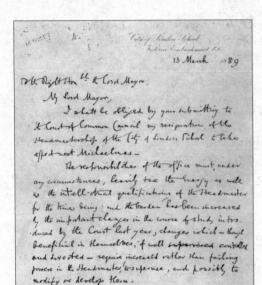

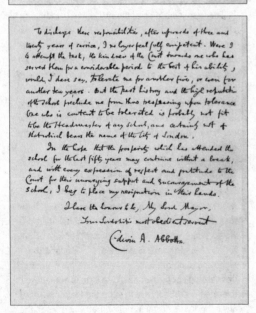

Figure 7 Abbott's letter of resignation.

consequently he ought (it is argued) to acknowledge that his countrymen are not only long and broad, but also (though doubtless in a very slight degree) *thick* or *high*. This objection is plausible, and, to Spacelanders, almost irresistible, so that, I confess, when I first heard it, I knew not what to reply. But my poor old friend's answer appears to me completely to meet it.

"I admit," said he—when I mentioned to him this objection—"I admit the truth of your critic's facts, but I deny his conclusions. It is true that we have really in Flatland a Third unrecognized Dimension called 'height,' just as it is also true that you have really in Spaceland a Fourth unrecognized Dimension, called by no name at present, but which I will call 'extra-height'. But we can no more take cognizance of our 'height' than you can of your 'extra-height'. Even I—who have been in Spaceland, and have had the privilege of understanding for twenty-four hours the meaning of 'height'—even I cannot now comprehend it, nor realize it by the sense of sight or by any process of reason; I can but apprehend it by faith.

"The reason is obvious. Dimension implies direction,[30] implies measurement, implies the more and the less. Now, all our lines are *equally* and *infinitesimally* thick (or high, whichever you like); consequently, there is nothing in them to lead our minds to the conception of that Dimension. No 'delicate micrometer'—as has been suggested by one too hasty Spaceland critic—would in the least avail us; for we should not know *what to measure, nor in what direction.* When we see a Line, we see something that is long and *bright; brightness,* as well as length, is necessary to the existence of a Line; if the brightness vanishes, the Line is extinguished. Hence, all my Flatland friends—when I talk to them about the unrecognized Dimension which is somehow visible in a Line—say, 'Ah, you mean *brightness*': and when I reply, 'No, I mean a real Dimension,' they at once retort, 'Then measure it, or tell us in what direction it extends'; and this silences me, for I can do neither. Only yesterday, when the Chief Circle (in other words our High Priest) came to inspect the State Prison and paid me his seventh annual visit, and when for the seventh time he put me the question, 'Was I any better?' I tried to prove to him that he was 'high,' as well as long and broad, although he did not know it. But what was his reply? 'You say I am "high"; measure my "high-ness" and I will believe you.' What could I do? How could I meet his challenge? I was crushed; and he left the room triumphant.

"Does this still seem strange to you? Then put yourself in a similar position. Suppose a person of the Fourth Dimension, condescending to visit you, were to say, 'Whenever you open your eyes, you see a Plane (which is of Two Dimensions) and you *infer* a Solid (which is of Three); but in reality you also see (though you do not recognize) a Fourth Dimension, which is not colour nor bright-

In the hope that the prosperity which has attended the school for the last fifty years may continue without a break, and with every expression of respect and gratitude to the Court for their unvarying support and encouragement of the school, I beg to place my resignation in their hands.

I have the honour to be, my Lord Mayor,
Your Lordship's most obedient servant,
Edwin A. Abbott.

L. R. Farnell says, "… Next to teaching, Abbott's vocation lay in writing; and it was probably the attraction of complete leisure for literary work, as well as his weariness of administration, which prompted his retirement at the zenith of his reputation.…" One major source of weariness with administration seems clear: During his tenure as headmaster, Abbott devoted a huge amount of time and energy to the authorization and construction of a new school (inaugurated on 23 January 1883), the old one having become inadequate (see the introduction to this book). As early as 1880 he wrote, "I should like to live long enough to see the boys into [the new school]. Then I shall feel quite ready to say Nunc Dimittis — at all events as regards dismission from the School into other provinces of work." (Nunc Dimittis is the Latin opening of the phrase "now lettest thou thy servant depart in peace.") A. E. Douglas-Smith's *City of London School* tells us that "… as far back as 1880 Abbott was looking forward to laying down his charge; not because he looked for rest, but because another call was always sounding in his ears."

29 The "correct" answer to this objection is that in a two-dimensional universe, the "eye" receiving light is also two-dimensional and has a one-dimensional retina. Therefore, light rays emitted from a one-dimensional object can successfully "map" the object onto the retina (Figure 8). Thus it is *not* necessary for the object to have "thickness" as well as length. Abbott is clearly aware of this, but he has the same problem in explaining it that stimulated the writing of

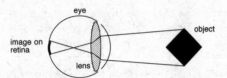

Figure 8 Mapping object to retina in two dimensions. No "thickness" in the third dimension is necessary.

Flatland in the first place: his readers' innate prejudice for three-dimensional ways of thinking. He solves his problem in the same manner: by making literal use of the analogy between the two dimensions of Flatland and the three of space. If the argument about thickness in Flatland is correct, then *by the same token,* what we see in three-dimensional space must actually have a slight but nonzero thickness along a fourth dimension — "extra-height." Abbott cleverly leaves it to his reader to absorb the point, which is not that "extra-height" must exist in space but that "thickness" is *not* required for visibility in Flatland.

30 The central scientific theme of *Flatland* is the meaning of dimension and its physical implications, and later notes will be more comprehensible if we take a few moments to review the concept. In modern mathematics there are many notions of dimension, each relevant to a different context. In Abbott's time only one notion had common currency, the idea that the dimension of a space (or object) is the number of *independent directions* that are available within it. Thus a line is one-dimensional because only one direction is available: along the line. For definiteness, assume the line runs from west to east. Motion to the west is merely the negative of motion to the east and does not count as a different direction. In the plane, however, there is a second direction, running from south to north. All other directions are combinations of these two: For example, northeast is an equal mixture of east and north. In space there is a third independent direction: up-down. Our physical space does not seem to

ness nor anything of the kind, but a true Dimension, although I cannot point out to you its direction, nor can you possibly measure it.' What would you say to such a visitor? Would not you have him locked up? Well, that is my fate: and it is as natural for us Flatlanders to lock up a Square for preaching the Third Dimension, as it is for you Spacelanders to lock up a Cube for preaching the Fourth. Alas, how strong a family likeness runs through blind and persecuting humanity in all Dimensions! Points, Lines, Squares, Cubes, Extra-Cubes — we are all liable to the same errors, all alike the Slaves of our respective Dimensional prejudices, as one of your Spaceland poets has said —[31]

'One touch of Nature makes all worlds akin'.[1]

On this point the defence of the Square seems to me to be impregnable. I wish I could say that his answer to the second (or moral) objection was equally clear and cogent. It has been objected that he is a woman-hater;[32] and as this objection has been vehemently urged by those whom Nature's decree has constituted the somewhat larger half of the Spaceland race, I should like to remove it, so far as I can honestly do so. But the Square is so

---

[1] The Author desires me to add, that the misconception of some of his critics on this matter has induced him to insert (on pp. 70 and 88) in his dialogue[33] with the Sphere, certain remarks which have a bearing on the point in question, and which he had previously omitted as being tedious and unnecessary.

unaccustomed to the use of the moral terminology of Spaceland that I should be doing him an injustice if I were literally to transcribe his defence against this charge. Acting, therefore, as his interpreter and summarizer, I gather that in the course of an imprisonment of seven years he has himself modified his own personal views, both as regards Women and as regards the Isosceles or Lower Classes. Personally, he now inclines to the opinion of the Sphere (see page 159) that the Straight Lines are in many important respects superior to the Circles. But, writing as a Historian, he has identified himself (perhaps too closely) with the views generally adopted by Flatland, and (as he has been informed) even by Spaceland, Historians; in whose pages (until very recent times) the destinies of Women and of the masses of mankind have seldom been deemed worthy of mention and never of careful consideration.

In a still more obscure passage he now desires to disavow the Circular or aristocratic tendencies with which some critics have naturally credited him. While doing justice to the intellectual power with which a few Circles have for many generations maintained their supremacy over immense multitudes of their countrymen, he believes that the facts of Flatland, speaking for themselves without comment on his part, declare that Revolutions cannot always be suppressed by slaughter, and that Nature, in sentencing the Circles to infecundity, has condemned them to ultimate failure

possess a fourth independent direction, but mathematically there is no barrier to spaces having as many dimensions as we wish, and, as C. H. Hinton urged in "What Is the Fourth Dimension?," there is value in "questioning whatever seems arbitrary and irrationally limited in the domain of knowledge."

A space can also have *no* dimensions — a point is an example. Abbott introduces Pointland in Chapter 20 of *Flatland* to illustrate complacency. The cover of *Flatland* (which was essentially the same in the first edition as in the Blackwell reissue of 1926; only the publisher, price, and paper quality have changed) depicts Pointland, Lineland, Flatland, and Spaceland and identifies them as having no, one, two, and three dimensions, respectively. Hovering enigmatically amid clouds, often partially obscured, are the phrases *Four dimensions* up to *Ten dimensions*.

31 The quotation is from Shakespeare's *Troilus and Cressida* (act 3, scene 3, line 175), where Ulysses, in discussion with Achilles, says, "One touch of Nature makes the whole world kin." Abbott's misquotation is presumably deliberate, to fit his context — comparing "Dimensional prejudices" in different worlds. As a Shakespeare scholar, he would have been aware of the exact original.

32 Authors who indulge in irony run the risk of being taken literally; authors who put opinions in the mouths of their characters, for whatever literary reason, run the risk of being identified with those opinions. The preface to the second edition of *Flatland* (published *one month* after the first, in 1884) makes it clear that Abbott must have found himself on the receiving end of criticism about the regard in which *Flatland* held its women. Characteristically, instead of being sarcastic about his critics' lack of perceptiveness, Abbott gently explains that A. Square "has himself modified his own personal views" — with regard not only to women but also to the Victorian class structure that *Flatland* parodies. Abbott's response to misunderstandings of the social irony in his first edition is another dose of heavy irony, presented as an apparent backing down by his central character. Abbott himself was a keen and active reformer who argued strongly for improvements in the education of women. As Banchoff tells us, he

> was a firm believer in equality of educational opportunity, across social classes and in particular for women. He participated actively in the efforts to bring about changes....[M]any of the women who gained entrance to universities, like Abbott's daughter, had received much of their education at home, often from private tutors....Abbott was also a vocal leader in the Teachers' Training Syndicate, formed and primarily supported by the major female educators of Victorian England, who extensively praised Abbott for his efforts on behalf of education reform.

33 These insertions are described in the appropriate places, below.

34 *Flatland* is primarily a fantasy, not "hard science fiction," and as such it makes no attempt to be logically consistent in every respect. However, it comes close enough to such consistency that its occasional logical flaws (and, no doubt, supposed flaws occasioned by readers' misunderstandings) must have been the subject of comment or correspondence. Logically more rigorous approaches to the science and sociology of a two-dimensional world can be found in *An Episode of Flatland* by C. H. Hinton, in which Flatland is replaced by a disk-shaped planet, Astria; in the 1957 *Bolland (Sphereland)* by the Dutch physicist Dionys Burger, whose narrative begins with Flatland and culminates in curved space-time; and in the Canadian computer scientist Alexander Keewatin ("Kee") Dewdney's *The Planiverse,* which relates the adventures of a creature named Yendred on the two-dimensional planet Arde. In *A Plane World* (1884), Charles Hinton lays the foundations for his later novel

—"and herein," he says, "I see a fulfilment of the great Law of all worlds, that while the wisdom of Man thinks it is working one thing, the wisdom of Nature constrains it to work another, and quite a different and far better thing." For the rest, he begs his readers not to suppose that every minute detail[34] in the daily life of Flatland must needs correspond to some other detail in Spaceland; and yet he hopes that, taken as a whole, his work may prove suggestive as well as amusing, to those Spacelanders of moderate and modest minds who —speaking of that which is of the highest importance, but lies beyond experience—decline to say on the one hand, "This can never be," and on the other hand, "It must needs be precisely thus, and we know all about it."

An Episode of Flatland (1907), introducing a world of two dimensions inhabited by human-like beings that he draws as triangles:

> Where the sun's rays grazing the earth in January pass off and merge into darkness lies a strange world.
>
> 'Tis a vast bubble blown in a substance something like glass, but harder far and untransparent.
>
> And just as a bubble blown by us consists of a distended film, so this bubble, vast beyond comparison, consists of a film distended and coherent.
>
> On its surface in the course of ages has fallen a thin layer of space dust, and so smooth is this surface that the dust slips over it to and fro and forms densities and clusters as its own attractions and movements determine.
>
> The dust is kept on the polished surface by the attraction of the vast film; but, except for that, it moves on it freely in every direction.
>
> And here and there are condensations wherein have fallen together numbers of these floating masses, and where the dust condensing for ages has formed vast disks.

It is on the circumference of such a disk that Hinton's beings live, and their world is not a plane but a nearly flat portion of the surface of a huge sphere:

> Those disks, though large, are so immeasurably small compared with the vast surface of the all-supporting bubble, that their movements seem to lie on a plane flat surface; the curving of the film on which they rest is so slight compared to their magnitude, that they sail round and round their central fires as on a perfect level surface.

In short, Hinton's beings live in a curved space, finite but unbounded, but to them it appears flat. Cosmologists now think that our own universe may well be like that. Unlike Abbott's polygons, Hinton's beings are confined to the rim of their circular planet, although they have limited movement in the up-down direction, just as we do. He goes to some lengths to think about what life in such a world would entail, in-

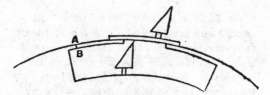

Figure 9 Hinton's solution to the problem of passing another inhabitant of the two-dimensional world Astria.

cluding special trapdoor arrangements for people to pass each other (Figure 9). One of the beings, Mulier (Latin for "woman"), is one day found in a mirror-image state. Later she disappears, eventually to be discovered during some excavations, sealed into an underground cavity. These baffling events have one simple explanation: She was moved through a third dimension.

In *A Picture of Our Universe* (1884) Hinton suggests the names *ana* and *kata* (Greek for "along" and "against," used by him in the sense "away from" and "toward") for the extra directions (akin to up-down) in a four-dimensional world. The article "Casting Out the Self" in his first volume of *Scientific Romances* (1884) describes a cube divided into 3 x 3 x 3 = 27 subcubes (Figure 10) and advocates their use for thinking about three-dimensional sections of a four-dimensional shape. *The Fourth Dimension* (1904) has an extended factual section

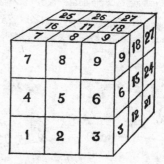

Figure 10 Hinton's diagram of a cube divided into 27 smaller cubes.

on non-Euclidean geometry. The short article "Recognition of the Fourth Dimension" (1902) speculates that "vital activities" might arise from molecules that are moving in the fourth dimension, changing from one form to its mirror image and back. As Rudy Rucker remarks in his edited collection of Hinton's writings, *Speculations of the Fourth Dimension,*

> The vitalist belief that life must involve some unusual physics is now pretty much discredited; and the 4-D rotation idea is definitely wrong, since it has been determined that our genetic material is in the form of a left-handed DNA helix which does not ever appear in the right-handed form.

Hinton's major work, in our context, is of course *An Episode of Flatland,* which develops a novelette-length story with proper characters and plot action, relegating its extra-dimensional speculations to the background. In the opening of the book, Hinton explains how he got the idea (and implicitly criticizes Abbott).

> Placing some coins on the table one day, I amused myself by pushing them about, and it struck me that one might represent a planetary system of a certain sort by their means. This large one in the centre represents the sun, and the others its planets journeying round it..... I saw that we must think of the beings that inhabit these worlds as standing out from the rims of them, not walking over the flat surface of them.

The story begins with the history of Astria, whose males all face east and its females west, and can never turn to face the other way. In the war between the western Unaeans and the eastern Scythians, this peculiarity of two-dimensional geometry offers a decisive advantage to the Unaeans (Figure 11), who sweep eastward and override their enemies. The plot revolves around the star-crossed lovers Laura Cartwright and Harold Wall. Laura's uncle, Hugh Farmer, has been convinced of the existence of the third dimension by seeing an item mysteriously vanishing from inside a closed box (compare folio 131). As mentioned in the introduction to this book, the plot involves a threat to Astria from the approaching planet Ardaea, which must be diverted. (The current interest in possible asteroid impacts on the Earth is nothing new in science fiction.) Farmer saves the world by persuading its inhabitants to lock on to a being that lives outside the Astrian plane, using a form of telekinesis to pull Astria out of the path of the rogue planet. To be fair, he offers a rationale for the telekinetic powers involved.

> Farmer's theory was two-fold, first, that by a grouping and rearranging of the molecular structure of the brain, such material changes could be effected as would cause a body slipping over the surface on which all Astrian things moved to be deflected in its course; and, secondly, that by thinking certain thoughts such molecular changes were produced in the brain matter of the thinkers.

After much debate with the priesthood and various acts of resistance, the idea is implemented as mass prayer, and the world is saved. Farmer then retires from the intellectual fray.

> He left the busy city and the crowds of men and, half mortified, half amused but wholly glad for now the danger to the dear world was over; he devoted himself to his garden in far away Scythia. That rebellious and antagonistic mind forgot its struggles and vicissitudes in watching the little beads of verdure that sprang out of the dark earth.

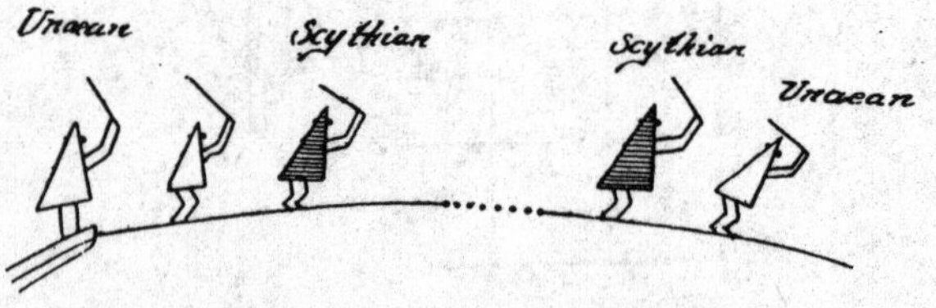

Figure 11 Why the Unaeans had a decisive advantage in battle.

Contents

35 In the first edition, this is "14. How in my Vision I endeavoured to explain the nature of Flatland, but could not."

PART I

THIS WORLD

"Be patient, for the world is broad and wide."[1]

1 The quotation is from Shakespeare's *Romeo and Juliet* (act 3, scene 3, line 16). It is said by Friar Lawrence, who has just told Romeo that he is to be banished from Verona. Romeo is not pleased because Verona is where the fair Juliet lives. This time the quotation (though not its Shakespearean meaning) is highly appropriate to Abbott's narrative: The world (Flatland) possesses the two dimensions of breadth and width — and, by implication, no others.

§ 1.—Of the Nature of Flatland.

I CALL our world Flatland,[1] not because we call it so, but to make its nature clearer to you, my happy readers, who are privileged to live in Space.

Imagine a vast sheet of paper[2] on which straight Lines, Triangles, Squares, Pentagons, Hexagons, and other figures, instead of remaining fixed in their places, move freely about, on or in the surface, but without the power of rising above or sinking below it, very much like shadows—only hard and with luminous edges—and you will then have a pretty correct notion of my country and countrymen. Alas, a few years ago, I should have said "my universe": but now my mind has been opened to higher views of things.

In such a country, you will perceive at once that it is impossible that there should be anything of what you call a "solid" kind;[3] but I dare say you will suppose that we could at least distinguish by sight the Triangles, Squares, and other figures, moving about as I have described them. On the contrary, we could see nothing of the kind, not at least so as to distinguish one figure from another. Nothing was visible, nor could be visible, to us, except Straight Lines; and the necessity of this I will speedily demonstrate.

1 With his opening sentence, Abbott transports his readers straight into a new universe. This is a science fiction type of opening, and ABBOTT, EDWIN A(BBOTT) is the second entry in *The Encyclopaedia of Science Fiction,* edited by John Clute and Peter Nicholls. Though not part of the classic science fiction genre — in part because it is too early — *Flatland* is widely admired in science fiction circles and belongs firmly to the prehistory of science fiction (henceforth SF).

Indeed, Flatland is one of the earliest works of what might be termed mathematics fiction — speculative fiction with a mathematical theme. In this respect it was preceded by *Alice's Adventures in Wonderland* (1865) and *Through the Looking-Glass and What Alice Found There* (1871) by Lewis Carroll (Charles Lutwidge Dodgson, 1832–1898). However, the mathematics in *Alice* is relatively well concealed: Dodgson was a mathematician in his "day job," and some of it rubs off in the narrative. In his poem *The Hunting of the Snark* (1876), the mathematical influence is more overt; see Martin Gardner's *The Annotated Snark.* Dodgson and Abbott have some things in common: Both were clergymen, both enjoyed mathematics, both loved writing, and they lived at much the same time. The differences between them are much greater than the similarities, however. For example, Abbott was a great teacher, whereas Dodgson's lectures were exceedingly dull and ordinary. As far as I can discern, there is no detectable Carrollian influence in *Flatland.* An even earlier example of "mathematics fiction" is the Laputa chapter of *Gulliver's Travels* (originally *Travels Into Several Remote Nations of the World in Four Parts … by Lemuel Gulliver,* 1726, revised 1735) by Jonathan Swift (1667–1745), which lampoons intellectuals of all kinds and mathematicians in particular.

The best-known *Flatland* derivative is the 1957 *Bolland: Een roman van gekromde ruimten en uitdijend heelal* (translated in 1965 as *Sphereland: a fantasy about curved spaces and an expanded universe*), in which the Dutch physicist Dionys (anglicized from Dionijs) Burger develops Flatland into a vehicle for explaining Einstein's ideas on the curvature of space-time. Dewdney's *The Planiverse* has a more sweeping aim: to explore the physics, chemistry, engineering, biology, sociology, and politics of a two-dimensional world. Like Hinton's Astria, Dewdney's planet Arde (a name suspiciously similar to Hinton's "Ardaea") is a disk, and the inhabitants occupy its surface. Yendred, an intelligent Ardean, makes contact with Earth when the computer program 2DWORLD takes on a life of its own. The story is complex, and the book is heavily illustrated with drawings on Ardean lifeforms and technology: a fishing-boat, a factory, even an internal combustion engine. An appendix discusses two-dimensional physics, chemistry, planetary science, biology, astronomy, and technology.

American journalist Martin Gardner (1914–), author from 1956 to 1981 of *Scientific American*'s celebrated Mathematical Games column, wrote several SF short stories based on mathematical ideas, including "The No-sided Professor." Symbolic logic is invoked in *The Incompleat Enchanter* (1942) and its sequels, by L(yon) Sprague de Camp (1907–?) and (Murray) Fletcher Pratt (1897–1956), as a way to transmit the protagonists into various fantasy worlds, such as that of the Norse gods or the *Faerie Queene* (1590) of Edmund Spenser (1552/3–1599). The stories of Norman Kagan involve hyper-active mathematics students, and the short story "The Mathenauts" (1964) (reprinted in, for example, *10th Annual SF* edited by Judith Merrill, 1965) uses an "isomorphomechanism" to transport ships into exotic mathematical spaces. The American mathematician Rudy Rucker (1946–) edited an anthology of mathematics fiction stories, *Mathenauts* (1987), which includes his own "Message Found in a Copy of *Flatland*". In *Neverness* (1988) and its sequels, by David Zindell (1952–), mathematics acquires a romantic aspect: The space-pilots in the Order of Mystic Mathematicians move at will through "windows" in the universe by proving theorems, and the

Place a penny on the middle of one of your tables in Space; and leaning over it, look down upon it. It will appear a circle.

But now, drawing back to the edge of the table, gradually lower your eye (thus bringing yourself more and more into the condition of the inhabitants of Flatland), and you will find the penny becoming more and more oval to your view; and at last when you have placed your eye exactly on the edge of the table[4] (so that you are, as it were, actually a Flatlander) the penny will then have ceased to appear oval at all, and will have become, so far as you can see, a straight line.

The same thing would happen if you were to treat in the same way a Triangle, or Square, or any other figure cut out of pasteboard. As soon as you look at it with your eye on the edge on the table, you will find that it ceases to appear to you a figure, and that it becomes in appearance a straight line. Take for example an equilateral Triangle—who represents with us a Tradesman of the respectable class. Fig. 1 represents the Tradesman as you would see him while you were bending over him from above; figs. 2 and 3 represent the Tradesman, as you would see him if your eye were close to the level, or all but on the level of the table; and if your eye were quite on the level of the table (and that

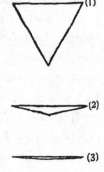

is how we see him in Flatland) you would see nothing but a straight line.

When I was in Spaceland I heard that your sailors have very similar experiences while they traverse your seas and discern some distant island or coast lying on the horizon. The far-off land may have bays, forelands, angles in and out to any number and extent; yet at a distance you see none of these (unless indeed your sun shines bright upon them revealing the projections and retirements by means of light and shade), nothing but a grey unbroken line upon the water.

Well, that is just what we see when one of our triangular or other acquaintances comes towards us in Flatland. As there is neither sun with us, nor any light of such a kind as to make shadows, we have none of the helps to the sight that you have in Spaceland. If our friend comes closer to us we see his line becomes larger; if he leaves us it becomes smaller: but still he looks like a straight line; be he a Triangle, Square, Pentagon, Hexagon, Circle, what you will—a straight Line he looks and nothing else.

You may perhaps ask how under these disadvantageous circumstances we are able to distinguish our friends from one another: but the answer to this very natural question will be more fitly and easily given when I come to describe the inhabitants of Flatland. For the present let me defer this subject, and say a word or two about the climate and houses in our country.

intellectual excitement the author conveys when describing these techniques is almost sexual in its intensity.

Flatland is the first example in SF of a two-dimensional universe, but numerous "flat" worlds have been invented since. The great classic is *Mission of Gravity* (1953) by the American science teacher Hal Clement (Harry Clement Stubbs, 1922–). The planet Mesklin has very high gravity but is lens-shaped because of its rapid rotation; gravity varies from 700g at the poles to 3g at the equator (1g = normal Earth gravity). Its hero, the alien Captain Barlennan, and his crew help to recover a vital component of a crashed Terran (SF for "from Earth") space probe. The two-dimensionality here is not the shape of Mesklin, but the result of its high gravity: Its centipede-like inhabitants are confined almost completely to its surface because a fall of even half an inch (1 cm) in 700g is inevitably fatal. The dénouement, as in *Flatland,* involves breaking through into the third dimension — in this case, by being told how to make a hot air balloon.

The same theme is taken to serious extremes by the American physicist Robert L(ull) Forward (1932–) in *Dragon's Egg,* set on the surface of a neutron star. Here the force of gravity is 67 *billion* g. The star is inhabited by the alien cheela, who evolve at a rate of one generation every 37 minutes. When humans encounter the primitive cheela, they accidentally civilize them during one 24-hour period. In the sequel *Starquake!* (1985), the cheela out-evolve humans and explore the entire galaxy.

A different "flat" world is the Discworld, scene of (so far) twenty-six humorous fantasy novels and numerous spin-offs by the immensely popular British writer Terry Pratchett (1948–). The Discworld is flat, circular, and about 10,000 miles (16,000 kilometers) in diameter. It is supported by four giant elephants who stand on the back of the great turtle A'Tuin, who swims through space on some incomprehensible cosmic journey. Discworld is populated by wizards, witches, elves, trolls, dwarves, and a variety of other creatures, including the talking dog Gaspode (no one notices he's talking because everyone knows dogs can't talk) and Cohen the Barbarian. Magic works on Discworld, and everything runs on narrative imperative — the power of story. Originally intended as a parody of generic fantasy and sword-and-sorcery novels, the Discworld canon quickly mutated into a parody of *everything:* feminism (*Equal Rites,* 1987), death (*Mort,* 1987, and *Reaper Man,* 1991), religion (*Small Gods,* 1992), opera (*Maskerade,* 1995), Australia (*The Last Continent,* 1998), and journalism (*The Truth,* 2000). Among the spin-offs is the fact/fantasy fusion *The Science of Discworld* (1999).

2 This is Abbott the schoolmaster, with an image straight out of the pages (literally) of Euclid. In Victorian times, "geometry" in school-level education followed closely the simpler parts of Euclid's *Elements.* Anyone who had taken geometry lessons at school would instantly be familiar with the flat world of a sheet of paper, populated by geometric figures. The only extra leap of imagination required was to animate those figures and endow them with human characteristics. The archives of the City of London School contain two old geometry texts. One is J. R. Young's *Euclid's Elements, Chiefly from the Texts of Simson and Playfair, with Corrections* (Souter, London 1839); the other is *Euclidian Geometry* by Francis Cuthbertson (Macmillan, London 1874). A note by Terry Heard, former head of mathematics at the City of London School, states that Young's book was used to teach geometry in the school around 1840, and the second was written by a former second master of the school. Indeed, Cuthbertson held that post from 1856 until his death in 1889 and was by all accounts a thoroughly normal and decent man, unlike his predecessor, the clever but crazy Edkins. Euclid's Proposition 46: *On a given straight line to describe a square,* also appears as Proposition 46 in Young's book (which shows just how closely the textbooks followed Euclid), but it has become Prob-

lem K in Cuthbertson's. The proofs offered in both texts are very similar to Euclid's.

3 On one level this statement seems so obvious that it needs no further thought, but mathematicians learned long ago to be suspicious of things that seem obvious. Abbott is saying that three-dimensional figures cannot exist inside a two-dimensional plane. This is true, but its proof is far from easy. The difficulty is compounded because there are many mathematical concepts of the "dimension" of a space, which depend on what additional structure the space is assumed to possess. The counterintuitive aspects of dimension can be illustrated by the space-filling curve of the Italian mathematician Giuseppe Peano (1858–1932) (Figure 12a). This is a continuous curve, which intuitively ought to be one-dimensional, but it completely fills the interior of a square, implying that it is actually two-dimensional. Another such curve was invented by the German mathematician David Hilbert (1862–1943) (Figure 12b). Similar constructions produce a curve that fills a cube (three dimensions) (Figure 12c) or indeed a "hypercube" with any finite number of dimensions.

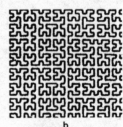
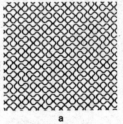
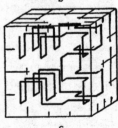

a

b

Figure 12 (a) Peano's space-filling curve in the plane. **(b)** Hilbert's space-filling curve in the plane. **(c)** Hilbert's space-filling curve in three dimensions. The curves shown are early stages in the contructions: the actual curves are defined by repeatedly adding ever smaller wiggles and continuing to the infinite limit.

c

4 If a triangle (or other connected two-dimensional plane figure) is projected *parallel* to the containing plane, the result is a line segment. Projection in any other direction preserves the dimension and results in a distorted two-dimensional shape. In fractal geometry, it can be proved that "almost all" projections preserve the dimension of a figure; that is, projections that do not preserve dimension satisfy a special condition that is rarely valid (such as "being parallel to the plane of the figure"). On the other hand, projections in different directions are essentially independent of each other. One consequence of this independence is the existence of a "digital sundial": a mathematical shape whose shadow on the ground, as the sun moves across the sky, tells the time *in digital form.* That is, at 3.26 its shadow looks like the numerals 3.26, at 3.27 its shadow looks like the numerals 3.27, and so on.

1 The basis of (rectilinear) Cartesian coordinates in the plane is the choice of two lines meeting at right angles, the axes of coordinates. Straight lines parallel to these form a "coordinate grid" of horizontal and vertical lines, which cross at right angles where they meet. On the surface of the Earth, there is a similar (but different) coordinate system: lines of latitude and longitude. The difference is that lines of latitude and longitude are circles, not straight lines, as becomes especially apparent near the poles. However, small regions of the Earth's surface are nearly flat, and this permits mapmakers to represent the local geography on a flat sheet of paper with a rectilinear coordinate grid. Conventionally, one coordinate axis runs from west to east and one from north to south. Thus Abbott is equipping Flatland with a Cartesian coordinate system analogous to what would be seen on a reasonably small map, such as one of the British Isles. His readers would be especially familiar with such maps because they were routinely produced by the Ordnance Survey, which was initiated in 1791 and published its first map (of the county of Kent) in 1801 on a scale of one inch to the mile. In 1858 this scale was approved by a royal commission for mapping the whole of Britain. In Flatland, however, the grid goes on forever: The territory is an infinite plane, not a sphere.

2 By making the north-south direction distinguishable from the east-west direction, Abbott is making the physics of Flatland anisotropic—that is, direction-dependent. However, he tells us that in the more temperate regions, this distinction becomes negligible and all directions appear essentially identical: Now the physics is isotropic—independent of direction. In modern mathematics these ideas are related to the *symmetries* of the plane—the "rigid motions" that preserve distance. There are three types of rigid motion: *translation* (in which the plane *slides* in some direction), *rotation* (in which it *turns* about some fixed center), and *reflection* (in which it *flips* over as though reflected in some mirror). See

§ 2.—*Of the Climate and Houses in Flatland.*

As WITH YOU, so also with us, there are four points of the compass[1]—North, South, East, and West.

There being no sun nor other heavenly bodies, it is impossible for us to determine the North in the usual way; but we have a method of our own. By a Law of Nature with us, there is a constant attraction to the South; and, although in temperate climates this is very slight—so that even a Woman in reasonable health can journey several furlongs northward without much difficulty—yet the hampering effect of the southward attraction is quite sufficient to serve as a compass in most parts of our earth. Moreover, the rain (which falls at stated intervals) coming always from the North, is an additional assistance; and in the towns we have the guidance of the houses, which of course have their side-walls running for the most part North and South, so that the roofs may keep off the rain from the North. In the country, where there are no houses, the trunks of the trees serve as some sort of guide. Altogether, we have not so much difficulty as might be expected in determining our bearings.

Yet in our more temperate regions,[2] in which

the southward attraction is hardly felt, walking sometimes in a perfectly desolate plain where there have been no houses nor trees to guide me, I have been occasionally compelled to remain stationary for hours together, waiting till the rain came before continuing my journey. On the weak and aged, and especially on delicate Females, the force of attraction tells much more heavily than on the robust of the Male Sex, so that it is a point of breeding, if you meet a Lady in the street, always to give her the North side of the way—by no means an easy thing to do always at short notice when you are in rude health and in a climate where it is difficult to tell your North from your South.

Windows there are none in our houses: for the light comes to us alike in our homes[3] and out of them, by day and by night, equally at all times and in all places, whence we know not. It was in old days, with our learned men, an interesting and oft-investigated question, "What is the origin of light?" and the solution of it has been repeatedly attempted, with no other result than to crowd our lunatic asylums with the would-be solvers. Hence, after fruitless attempts to suppress such investigations indirectly by making them liable to a heavy tax, the Legislature, in comparatively recent times, absolutely prohibited them. I—alas, I alone in Flatland—know now only too well the true solution of this mysterious problem; but my knowledge cannot be made intelligible to a single one of

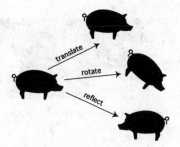

Figure 13 Three types of rigid motion in the plane.

Figure 13. The anisotropic plane of Flatland is symmetric (appears not to change) under any translation and reflection in any mirror that runs north-south. In the temperate regions, it also becomes symmetric under any rotation and any reflection. This difference affects the physics of Flatland because (as Einstein emphasized in our own universe) the laws of physics reflect the symmetries of the space(-time) in which they hold.

3 Flatland's light is like that of the schoolboy's lamp on his page from Euclid: It is generated by a source *outside* the plane. Strictly speaking, this assumption contains unnecessary traces of three-dimensional thinking; the light of a truly two-dimensional universe would not require an external source. A sufficiently perceptive Flatlander might deduce the existence of the third dimension from the properties of Flatland light as Abbott describes them—contrary to his central analogy, because light in *our* universe does not require an external source. However, Abbott is forced to make this assumption for plot reasons. If light in Flatland came from internal sources, every inhabitant would leave a long shadow (infinitely long for a point light source; see Figure 14). Flatland vision, and the entire narrative, would be unduly complicated by shadows. For example, the method for ascertaining the rank of an individual by visual means (page 66) would depend on where the light was coming from.

This is an elegant literary solution to a narrative

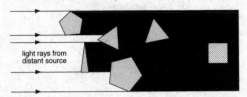

Figure 14 If there were shadows in Flatland, what would A. Square see?

problem, but to some extent the problem is occasioned by Abbott's decision to allow the Flatlanders to roam at will throughout their Euclidean plane. Humans (at least if we ignore the invention of flight) inhabit a thin region on the surface of an approximate sphere, a *two*-dimensional surface. Our light source comes from outside that surface: above during the day, tangential at dawn and dusk, and below (hence blocked by the Earth) at night. During the main part of the day, shadows do not block our visual perceptions, and we are in the same position as a Flatlander with an external light source. At night, we would see *only* a shadow (that of the Earth) were it not for starlight, moonlight, or artificial lighting. If Abbott had pursued his analogy more relentlessly, as did Hinton and Dewdney, he would have placed his creatures on the surface of a circular disk, with a distant external sun (also a disk), and light would have come from above during the day and would have been blocked by the disk-planet at night. Of course, such an arrangement leads to other narrative problems, such as how two creatures pass each other when they meet.

A modern disk-shaped world, the aforementioned Discworld of Terry Pratchett, handles light in quite a different manner. There, the disk is supported by four giant elephants standing on the back of a huge turtle, and the sun is a small, hot body about sixty miles away that orbits half-above, half-below the plane of the disk. (One of the elephants has to keep lifting a foot to let the sun pass.) On Discworld there are two main kinds of light: light you see *things* by and light by which you see the dark. See *The Sci-*

my countrymen; and I am mocked at—I, the sole possessor of the truths of Space and of the theory of the introduction of Light from the world of three Dimensions—as if I were the maddest of the mad! But a truce to these painful digressions: let me return to our houses.

The most common form for the construction of a house is five-sided or pentagonal, as in the annexed figure. The two Northern sides *RO, OF,*[4] constitute the roof, and for the most part have no doors; on the East is a small door for the Women; on the West a much larger one for the Men; the South side or floor is usually doorless.

Square and triangular houses are not allowed, and for this reason. The angles of a Square[5] (and still more those of an equilateral Triangle,) being much more pointed than those of a Pentagon, and the lines of inanimate objects (such as houses) being dimmer than the lines of Men and Women, it follows that there is no

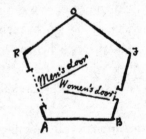

little danger lest the points of a square or triangular house residence might do serious injury to an inconsiderate or perhaps absentminded traveller suddenly running against them: and therefore, as early as the eleventh century of our era, triangular houses were universally forbidden by Law, the only exceptions being fortifications,[6] powder-magazines, barracks, and other state buildings, which it is not desirable that the general public should approach without circumspection.

At this period, square houses were still everywhere permitted, though discouraged by a special tax. But, about three centuries afterwards, the Law decided that in all towns containing a population above ten thousand, the angle of a Pentagon was the smallest house-angle that could be allowed consistently with the public safety. The good sense of the community has seconded the efforts of the Legislature; and now, even in the country, the pentagonal construction has superseded every other. It is only now and then in some very remote and backward agricultural district that an antiquarian may still discover a square house.

ence of Discworld for a discussion of how this is related to "privatives": concepts defined as the absence of something else.

4 Observe the visual pun in which the letters R, O, and F, designating three vertices of the pentagonal house, spell out *roof* when the two adjacent sides are specified — using Euclid's convention that the line joining a point P to a point Q should be named PQ. This is the only such visual pun in *Flatland,* and indeed it seems to be the only pun. Abbott had a sense of humor but reserved it for appropriate occasions. He instituted regular "Declamations" at his school, in which each boy composed and presented a speech on a topic of his own choice. The school has a legend that Robert Chalmers (1858–1938, later a high-ranking civil servant and first Baron Chalmers of Northiam) chose the topic of cremation and began by stating that it was "a grave subject and a burning question." Abbott immediately cut the performance short with a thunderous "Sit down, Chalmers, sit down!"

5 Here and elsewhere, Abbott alludes to features of Euclidean geometry that would be taught in most school texts. Ever the educator, he can't resist a gentle reminder to his readers. The internal angles of a regular n-sided polygon (or n-gon) are equal to $180-360/n$ degrees (Figure 15). For polygons with 3, 4, 5, 6 sides these angles are 60°, 90°, 108°,

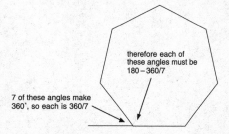

therefore each of these angles must be $180-360/7$

7 of these angles make 360°, so each is 360/7

Figure 15 Internal angles of a regular n-sided polygon: here we take $n=7$.

120° — the angle becomes larger as the number of sides increases. Small angles make sharp points; larger angles make blunter ones. As the number of sides becomes very large (as in the so-called Circles, the Flatland priesthood), the internal angle becomes very close to 180°, a straight line. This is as "blunt" as an angle can get.

6 Abbott is mostly having some fun here with the idea that sharp corners are good for keeping people away, but he was probably aware that in the seventeenth century, fortifications developed increasingly elaborate angular perimeters, often star-shaped (Figure 16). The reason, though, was not to achieve sharp edges. It was so that the bases of the walls of the fortification could be fired upon *from inside the fortification itself.* This was a good defense against tunnels or other methods for destroying the walls.

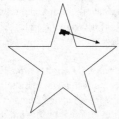

Figure 16 Star-shaped fortifications make it possible to defend the outside of the castle walls from inside the castle.

§ 3. — *Concerning the Inhabitants of Flatland.*

THE GREATEST length or breadth of a full grown inhabitant of Flatland may be estimated at about eleven of your inches.[1] Twelve inches may be regarded as a maximum.

Our Women are Straight Lines.[2]

Our Soldiers and Lowest Classes of Workmen are Triangles with two equal sides, each about eleven inches long, and a base or third side so short (often not exceeding half an inch) that they form at their vertice a very sharp and formidable angle. Indeed when their bases are of the most degraded type (not more than the eighth part of an inch in size), they can hardly be distinguished from Straight Lines or Women; so extremely pointed are their vertices. With us, as with you, these Triangles are distinguished from others by being called Isosceles; and by this name I shall refer to them in the following pages.

Our Middle Class consists of Equilateral or Equal-Sided Triangles.

Our Professional Men and Gentlemen are Squares (to which class I myself belong) and Five-Sided Figures or Pentagons.

Next above these come the Nobility, of whom there are several degrees, beginning at Six-Sided

1 This would make the typical size of a Flatlander about that of an engineering drawing or a "construction" by ruler and compasses in a geometry class. The standard school ruler in Abbott's day (as now) was 12 inches (30 centimeters) long. This places an upper limit on easily drawable lines.

2 The males of Flatland, even the most subservient and unintelligent, are two-dimensional figures. "Our Soldiers and Lowest Classes of Workmen are Triangles with two equal sides, each about eleven inches long, and a base or third side so short (often not exceeding half an inch)....Indeed when their bases are of the *most degraded type* [my italics], (not more than the eighth part of an inch in size), they can hardly be distinguished from Straight Lines or Women." Flatland's females are one-dimensional, lower in the social hierarchy than even a male of the most degraded type. In Flatland, one's social class is visible in one's physical form, with women right at the bottom of the heap.

In Spaceland, social class is also distinguished by visual cues: expensive clothes, jewelry, cars ... and (in Europe at least) by aural cues, a typical case being the classic "upper-class accent" of the English "public" schools. As usual, Abbott is accentuating features of Victorian society by turning them into physical attributes of his fictional creatures. As regards upper-class accents, *Nature* (volume 408, number 6815, 21/28 December 2000, page 927) reports a study of the Christmas messages of Queen Elizabeth II between the 1950s and the 1980s. The study, by Jonathan Harrington, Sallyanne Palethorpe, and Catherine Watson, bears the title "Does the Queen Speak the Queen's English?" They show that over that period, the vowel sounds of the Queen's pronunciation drifted about halfway from the original upper-class Queen's English of 1950 toward the 1980s accent of the southern commoner.

3 Abbott's mathematical care here is exemplary. A regular *n*-gon with *n* very large is approximately a circle, but the radius *r* of that circle depends on the length of side *d* of the polygon. In fact,

$$r = (d/2) \csc (\pi/n)$$

In order to approximate a given circle by a series of regular *n*-gons, for increasing *n*, the sides must shrink as *n* increases. Indeed, in the limit of *n* tending to infinity, for a circle of radius *r*, the side of the polygon must get closer and closer to $2\pi\, r/n$.

Approximation schemes of this kind were used by early mathematicians to find approximations to π (pi), the ratio of the circumference of a circle to its diameter. For example, in his *Measurement of a Circle*, Archimedes (287–212 B.C.) approximated the circle by inscribed and circumscribed 96-gons, proving that $3^{10}/_{71} < \pi < 3^{1}/_{7}$. Such approximations are unavoidable because the number π is *irrational*, meaning that it cannot be represented as an exact fraction. This fact was first proved in 1761 by Johann Heinrich Lambert (1728–1777). In 1882 Ferdinand Lindemann (1852–1939) proved a stronger result: π is transcendental. That is, it is not the solution of any polynomial equation with whole-number coefficients. This is why any numerical or algebraic expression for π must be approximate. The usual school approximation is $^{22}/_{7}$, accurate to two decimal places; a closer one is $^{355}/_{113}$, accurate to nine decimal places.

4 Abbott is describing an evolutionary process. In 1859 Charles Darwin (1809–1882) published *On the Origin of Species by Means of Natural Selection*. This epic work and its 1871 sequel *The Descent of Man and Selection in Relation to Sex* had a major impact on Victorian society, shaking it to its roots; the theory of evolution was very much in the air. We know that Abbott was interested in evolution because in 1897 he wrote *The Spirit on the Waters: The Evolution of the Divine from the Human*. Chapter 1, "The Evolution of Man," includes the following passages:

Figures, or Hexagons, and from thence rising in the number of their sides till they receive the honourable title of Polygonal, or many-sided. Finally when the number of the sides becomes so numerous, and the sides themselves so small,[3] that the figure cannot be distinguished from a circle, he is included in the Circular or Priestly order; and this is the highest class of all.

It is a Law of Nature with us[4] that a male child shall have one more side than his father, so that each generation shall rise (as a rule) one step in the scale of development and nobility. Thus the son of a Square is a Pentagon; the son of a Pentagon, a Hexagon; and so on.

But this rule applies not always to the Tradesmen, and still less often to the Soldiers, and to the Workmen; who indeed can hardly be said to deserve the name of human Figures, since they have not all their sides equal. With them therefore the Law of Nature does not hold; and the son of an Isosceles (*i.e.* a Triangle with two sides equal) remains Isosceles still. Nevertheless, all hope is not shut out, even from the Isosceles, that his posterity may ultimately rise above his degraded condition. For, after a long series of military successes, or diligent and skilful labours, it is generally found that the more intelligent among the Artisan and Soldier classes manifest a slight increase of their third side or base, and a shrinkage of the two other sides. Intermarriages (arranged by the Priests) between the sons and daughters of these more intellectual members of the lower classes generally result in an offspring approximating still more to the type of the Equal-Sided Triangle.

Rarely—in proportion to the vast numbers of Isosceles births—is a genuine and certifiable Equal-Sided Triangle[5] produced from Isosceles parents.[1] Such a birth requires, as its antecedents, not only a series of carefully arranged intermarriages, but also a long-continued exercise of frugality and self-control on the part of the would-be ancestors of the coming Equilateral, and a patient, systematic, and continuous development of the Isosceles intellect through many generations.

The birth of a True Equilateral Triangle from Isosceles parents is the subject of rejoicing in our country for many furlongs round. After a strict examination conducted by the Sanitary and Social Board, the infant, if certified as Regular, is with solemn ceremonial admitted into the class of Equilaterals. He is then immediately taken from his proud yet sorrowing parents and adopted by some childless Equilateral, who is bound by oath never to permit the child henceforth to enter his former home or so much as to look upon his relations again, for fear lest the freshly developed organism

[1] "What need of a certificate?" a Spaceland critic may ask: "Is not the procreation of a Square Son a certificate from Nature herself, proving the Equal-sidedness of the Father?" I reply that no Lady of any position will marry an uncertified Triangle. Square offspring has sometimes resulted from a slightly Irregular Triangle; but in almost every such case the Irregularity of the first generation is visited on the third; which either fails to attain the Pentagonal rank, or relapses to the Triangular.

The children of the vegetable world live, struggle for life with each other, and die. Many perish, few survive. ...In the animal but non-human world there arises a fiercer, though more restricted, struggle....The animal world seems divided into the preying and the preyed-on....Having fore-thought, man had before his mind the constant fear of being devoured by his superiors in bodily strength, the carnivora.

These are classic survival-of-the-fittest images, "Nature, red in tooth and claw" as Alfred, Lord Tennyson (1809–1892) wrote in his poem *In Memoriam*—which, interestingly, was written in 1850, eight years before Darwin published his theory but thirteen years after he first began groping toward it. The focus of *The Spirit on the Waters* is how evolution is related to human morality.

However, what Abbott is describing in this passage from *Flatland* is not Darwinian natural selection—not even its survival-of-the-fittest dumbing down—but the widespread Victorian misreading in which the central feature of evolutionary change is *progress* toward a higher state. This misreading appealed to the Victorians because it provided a biological justification for an inequitable society driven by privilege—the worst kind of "social Darwinism." It was *good* not to help the poor to improve their social condition because raw competition improved the human species. Abbott's instincts were more egalitarian than those of many of his contemporaries, though, and this showed in the attitudes of his school, as stated in *City of London School* by Douglas-Smith:

The School has ... always been progressive; it has also always been liberal and democratic. In the early nineteenth century it was almost impossible for Jews, Free Churchmen, and Roman Catholics to get a secondary education in England, except in their own schools. But the doors of C.L.S. were open from the first; the anti-Semitism of Fascist countries is incomprehensible to boys who have made their life's friendships in the School, and many of her most famous

sons have been Jews. No sooner had the School
Board of 1870 been established than Abbott, the
School's greatest Headmaster, founded a scholar-
ship for a Board school boy at C.L.S.

Later parts of *Flatland* cast doubt on the inevitability
of progress up the social scale.

5 In contrast to the previous paragraph, the image
here is closer to the now discredited theories of
Jean-Baptiste Lamarck (1744–1829) about the in-
heritance of acquired characters as expounded in
his *Philosophie Zoologique* (*Zoological Philosophy*)
of 1809. For example, a blacksmith becomes heavily
muscled by virtue of plying his trade, but according
to the Lamarckian view, this increases the likelihood
that his sons will also be heavily muscled.

6 It is a common feature of hierarchical societies
that the lower classes "know their place" and do not
strive to improve their station. In effect they become
accomplices in the maintenance of the hierarchy,
possibly because they know that any attempts to
overthrow it will lead to violent repression, but mostly
because life is much simpler if you know your place.
However, serfs can hope for improvements for their
children, and *very* occasionally, the higher classes
permit this (throwing a few scraps from the table to
the dogs) as a way of helping to keep the masses
quiet. The classic case in the United Kingdom is the
"working-class Tory" who repeatedly votes for the
party that supports privilege and thus helps maintain the status quo.

  This passage of the book is heavily laden with social satire, as is much of Part I of *Flatland,* which is
titled "This World." The more scientific Part II is titled
"Other Worlds." *This* world appears to refer to Flatland, but Abbott may have intended it more literally.

7 In Flatland, intelligence and social status are
closely associated, and intelligence can be measured directly in terms of a simple physical attribute:

may, by force of unconscious imitation, fall back
again into his hereditary level.

  The occasional emergence of an Equilateral from
the ranks of his serf-born ancestors is welcomed, not
only by the poor serfs themselves,[6] as a gleam of
light and hope shed upon the monotonous squalor
of their existence, but also by the Aristocracy at
large; for all the higher classes are well aware that
these rare phenomena, while they do little or nothing to vulgarize their own privileges, serve as a
most useful barrier against revolution from below.

  Had the acute-angled rabble been all, without
exception, absolutely destitute of hope and of
ambition, they might have found leaders in some
of their many seditious outbreaks, so able as to
render their superior numbers and strength too
much even for the wisdom of the Circles. But a
wise ordinance of Nature has decreed that, in
proportion as the working-classes increase in intelligence, knowledge, and all virtue,[7] in that same
proportion their acute angle (which makes them
physically terrible) shall increase also and approximate to the comparatively harmless angle of the
Equilateral Triangle. Thus, in the most brutal and
formidable of the soldier class—creatures almost
on a level with women in their lack of intelligence
—it is found that, as they wax in the mental ability necessary to employ their tremendous penetrating power to advantage, so do they wane in
the power of penetration itself.

  How admirable is this Law of Compensation!

And how perfect a proof of the natural fitness and, I may almost say, the divine origin of the aristocratic constitution of the States in Flatland! By a judicious use of this Law of Nature, the Polygons and Circles are almost always able to stifle sedition in its very cradle, taking advantage of the irrepressible and boundless hopefulness of the human mind. Art also comes to the aid of Law and Order. It is generally found possible — by a little artificial compression or expansion on the part of the State physicians — to make some of the more intelligent leaders of a rebellion perfectly Regular, and to admit them at once into the privileged classes; a much larger number, who are still below the standard, allured by the prospect of being ultimately ennobled, are induced to enter the State Hospitals, where they are kept in honourable confinement for life; one or two alone of the more obstinate, foolish, and hopelessly irregular are led to execution.

Then the wretched rabble of the Isosceles, planless and leaderless, are either transfixed without resistance by the small body of their brethren whom the Chief Circle keeps in pay for emergencies of this kind; or else more often, by means of jealousies and suspicions skilfully fomented among them by the Circular party, they are stirred to mutual warfare, and perish by one another's angles. No less than one hundred and twenty rebellions are recorded in our annals, besides minor outbreaks numbered at two hundred and thirty-five; and they have all ended thus.

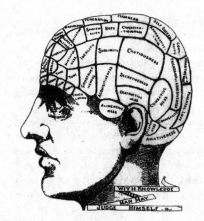

Figure 17 Phrenological regions of the brain.

the size of a Flatlander's angles. The bigger the angle, the bigger the brain; the bigger the brain, the greater the intelligence. Victorian society also assumed that people higher up the social scale were superior in every respect, including intellect. Most Victorians also believed that bigger brains implied superior intelligence. They were fascinated by the pseudoscience of phrenology, which attempted to relate personality to skull measurements. Phrenology was introduced by Franz-Joseph Gall (1758–1828) and flourished in the 1800s. It is based on five assumptions:

1. The mind resides in the brain.
2. Human mental abilities arise from a combination of a definite number of independent faculties.
3. Each such faculty is located in a specific region of the brain (Figure 17).
4. The size of that region in any individual determines how well developed that faculty is in that individual.
5. The external shape of the skull corresponds to that of the underlying regions of the brain, so it is possible to *measure* mental abilities from the outside (using, for example, calipers to "read the bumps").

Gall identified exactly twenty-six distinct "organs" of the brain, in specific positions. For instance, the "maths bump" or "sense of numerical relations" was placed just behind the eye.

The idea that the size of a brain determines its level of intelligence is a seductive but simplistic one; moreover, given that women are generally smaller than men and hence have smaller brains, it helps justify the conventional Victorian view — still widely believed even today, almost exclusively by men — that females are less intelligent than males. A leading advocate of this view in Victorian times was Paul Broca (1824–1880). When Broca died, his brain was removed from his skull and measured: It was found to be unusually small.

In modern science, mind (with the associated mental powers) is generally considered to be a consequence of the physical nature of the brain, in contrast to Descartes's belief that mind was a separate entity from the brain, made from entirely different stuff (Cartesian dualism). But what *kind* of consequence? Size seems relevant only in a qualitative manner: Too small a brain lacks the computing power to emulate the human one, but among human brains, size does not correlate meaningfully with mental ability. Even today, scientists often seek explanations of subtle mental differences by investigating the gross physiology of the brain. For example, Einstein's brain was preserved after his death, for scientific examination, and innumerable attempts have been made to observe some difference in its structure, compared to a normal brain, that might account for his unusual powers. Einstein's brain weighed about 1,200 grams (toward the low end of the scale). However, in 1995 it was discovered that he had a higher-than-average density of glial cells in an area of the brain known as the angular gyrus (or Brodmann's area 39), which is part of the lower left parietal lobe, a part of the brain that is associated with mathematical ability.

Unfortunately for this theory, Einstein himself stated that he found mathematics very difficult. (He just didn't let that stop him.) When developing the General Theory of Relativity, he spent several years attempting to master the tensor calculus of the Italian school of geometers and found it very hard going. In 1912 he wrote to his friend the mathematician Marcel Grossman (1878–1936), "Grossman, you must help me or else I'll go crazy!" Grossman suggested that he should take a look at the differential geometry of Georg Friedrich Bernhard Riemann (1826–1866; see "The Fourth Dimension in Mathematics" in this book), Gregorio Ricci (-Curbastro, 1853–1925), and Tullio Levi-Civita (1873–1941). There Einstein found what he needed, and in October 1912 he wrote to Arnold Sommerfeld (1868–1951) that

> At present I occupy myself exclusively with the problem of gravitation and now believe that I shall master all difficulties with the help of a friendly mathematician here. But one thing is certain, in all my life I have laboured not nearly as hard, and I have become imbued with great respect for mathematics, the subtler part of which I had in my simple-mindedness regarded as pure luxury until now. Compared with this problem, relativity is child's play.

No doubt *some* mental powers can be traced to specific physiological features of the brain, but most probably depend on subtle properties of how the brain cells are connected together.

§ 4. — Concerning the Women.

1 This is a reprise of the mathematical concept of projection, only now we observe that when a one-dimensional line is projected in certain special directions (parallel to the line), it reduces to a zero-dimensional point.

IF OUR HIGHLY pointed Triangles of the Soldier class are formidable, it may be readily inferred that far more formidable are our Women. For, if a Soldier is a wedge, a Woman is a needle; being, so to speak, all point, at least at the two extremities. Add to this the power of making herself practically invisible[1] at will, and you will perceive that a Female, in Flatland, is a creature by no means to be trifled with.

But here, perhaps, some of my younger Readers may ask *how* a woman in Flatland can make herself invisible. This ought, I think, to be apparent without any explanation. However, a few words will make it clear to the most unreflecting.

Place a needle on a table. Then, with your eye on the level of the table, look at it side-ways, and you see the whole length of it; but look at it endways, and you see nothing but a point, it has become practically invisible. Just so is it with one of our Women. When her side is turned towards us, we see her as a straight line; when the end containing her eye or mouth—for with us these two organs are identical—is the part that meets our eye, then we see nothing but a highly lustrous point; but when the back is presented to our view, then—being only sub-lustrous, and, indeed, almost as dim as an inanimate object—her hinder extremity serves her as a kind of Invisible Cap.

The dangers to which we are exposed from our Women must now be manifest to the meanest capacity in Spaceland. If even the angle of a respectable Triangle in the middle class is not without its dangers; if to run against a Working Man involves a gash; if collision with an Officer of the military class necessitates a serious wound; if a mere touch from the vertex of a Private Soldier brings with it danger of death;—what can it be to run against a Woman, except absolute and immediate destruction? And when a Woman is invisible, or visible only as a dim sub-lustrous point, how difficult must it be, even for the most cautious, always to avoid collision!

Many are the enactments made at different times in the different States of Flatland, in order to minimize this peril; and in the Southern and less temperate climates, where the force of gravitation is greater, and human beings more liable to casual and involuntary motions, the Laws concerning Women are naturally much more stringent. But a general view of the Code may be obtained from the following summary:—

1. Every house shall have one entrance in the Eastern side, for the use of Females only; by which all females shall enter "in a becoming and

respectful manner"[1] and not by the Men's or Western door.

2. No Female shall walk in any public place without continually keeping up her Peace-cry, under penalty of death.

3. Any Female, duly certified to be suffering from St. Vitus's Dance, fits, chronic cold accompanied by violent sneezing, or any disease necessitating involuntary motions, shall be instantly destroyed.

In some of the States there is an additional Law forbidding Females, under penalty of death, from walking or standing in any public place without moving their backs constantly from right to left so as to indicate their presence to those behind them; others oblige a Woman, when travelling, to be followed by one of her sons, or servants, or by her husband; others confine Women altogether to their houses except during the religious festivals. But it has been found by the wisest of our Circles or Statesmen that the multiplication of restrictions on Females tends not only to the debilitation and diminution of the race, but also to the increase of domestic murders to such an extent that a State loses more than it gains by a too prohibitive Code.

[1] When I was in Spaceland I understood that some of your Priestly circles have in the same way a separate entrance for Villagers, Farmers and Teachers of Board Schools (*Spectator*, Sept. 1884, p. 1255) that they may "approach in a becoming and respectful manner."[2]

2 This footnote, which brings the reader back to reality with a bump and rather spoils the illusion created by the narrative, is already present in the first edition, so it is not — as might seem plausible — a response to critics. It appears to be a pre-emptive strike.

3 At this stage in the narrative, the Flatland women begin to display signs of intelligence (and the same occurs in several later passages). The alert reader will realize that the true abilities of Flatland's women are potentially greater than those permitted by the role that society has allocated to them, a theme I develop in *Flatterland*. Abbott believed that the conventional roles of women in Victorian society were needlessly limiting, as shown by his interest in improving the education of women, which in his day was confined to traditional domestic accomplishments.

For whenever the temper of the Women is thus exasperated[3] by confinement at home or hampering regulations abroad, they are apt to vent their spleen upon their husbands and children; and in the less temperate climates the whole male population of a village has been sometimes destroyed in one or two hours of simultaneous female outbreak. Hence the Three Laws, mentioned above, suffice for the better regulated States, and may be accepted as a rough exemplification of our Female Code.

After all, our principal safeguard is found, not in Legislature, but in the interests of the Women themselves. For, although they can inflict instantaneous death by a retrograde movement, yet unless they can at once disengage their stinging extremity from the struggling body of their victim, their own frail bodies are liable to be shattered.

The power of Fashion is also on our side. I pointed out that in some less civilized States no female is suffered to stand in any public place without swaying her back from right to left. This practice has been universal among ladies of any pretensions to breeding in all well-governed States, as far back as the memory of Figures can reach. It is considered a disgrace to any State that legislation should have to enforce what ought to be, and is in every respectable female, a natural instinct. The rhythmical and, if I may so say, well-modulated undulation of the back in our ladies of Circular rank is envied and imitated by the wife of a common Equilateral, who can achieve nothing beyond a mere monotonous swing, like the ticking of a pendulum; and the regular tick of the Equilateral is no less admired and copied by the wife of the progressive and aspiring Isosceles, in the females of whose family no "back-motion" of any kind has become as yet a necessity of life. Hence, in every family of position and consideration, "back motion" is as prevalent as time itself; and the husbands and sons in these households enjoy immunity at least from invisible attacks.

Not that it must be for a moment supposed that our Women are destitute of affection. But unfortunately the passion of the moment predominates, in the Frail Sex, over every other consideration. This is, of course, a necessity arising from their unfortunate conformation. For as they have no pretensions to an angle, being inferior in this respect to the very lowest of the Isosceles, they are consequently wholly devoid of brain-power,[4] and have neither reflection, judgment nor forethought, and hardly any memory. Hence, in their fits of fury, they remember no claims and recognize no distinctions. I have actually known a case where a Woman has exterminated her whole household, and half an hour afterwards, when her rage was over and the fragments swept away, has asked what has become of her husband and her children.

Obviously then a Woman is not to be irritated as long as she is in a position where she can turn round. When you have them in their apartments

4 How would a Flatlander's brain work? Wiring up a two-dimensional brain is not straightforward because in two dimensions, there is no room for wires to cross each other without intersecting. However, the mathematical theory of *cellular automata* shows that two-dimensional arrays of cells, communicating with their neighbors according to simple rules, can carry out any task that a computer can accomplish, so a Flatlander might have a brain based on a cellular automaton.

Cellular automata were invented by the mathematician John von Neumann (1903–1957) in the 1950s, in a successful (but just too late) attempt to establish the feasibility of self-replicating machines. (At that time, many biologists considered replication to be beyond the powers of the physical world.) He did this in an abstract setting: a very large grid of squares in the plane — about 200,000 of them — arranged like the cells of a chessboard. At any instant each cell could exist in one of a large but finite number (he used 29) of "internal states." As the clock ticked, each cell would change state according to specific rules, based on the states of its neighbors. It is customary to use colors to designate the states; a typical rule might be "Any red cell adjacent to two blue cells and two green ones must turn yellow." The big technical problem that von Neumann had to overcome had long been seen as the basis of an apparently conclusive philosophical objection to the whole idea. If an object is to copy itself, then the copying instructions, which must be part of the object, must nevertheless specify the whole object — and then that specification must also be copyable, so the copying instructions must specify the specification … leading to an apparent infinite regress. Von Neumann solved the problem by splitting the object into three parts:

- A builder, which can build pretty much anything, given appropriate instructions
- A copier, which can copy any list of instructions
- The list of instructions needed to make the builder and the copier

Then the cellular automaton can reproduce itself in a series of stages:

- Give the builder the instruction list, so that it makes a new builder and a new copier and bolts them together.
- Give the copier the instruction list, so that it makes a new copy of *that*.
- Get the builder to tack on the copy of the instruction list in the appropriate place.

The essential trick is that the instruction list has two distinct interpretations: as a meaningless list, to be copied, and as a series of meaningful instructions, to be obeyed. This dual interpretation avoids the infinite regress. Von Neumann managed to specify an abstract self-copying cellular automaton, along these lines. But before he could publish, Francis Crick (1916–) and James Watson (1928–) worked out the molecular structure of DNA, and it quickly became clear that living cells replicate in the same abstract manner, with the cell as the builder, a system of enzymes as the copier, and DNA as the instructions. Von Neumann's work was robbed of what might otherwise have been a major impact: a mathematical prediction about the mechanism of replication in living creatures.

The idea of a cellular automaton didn't completely go away, though, and around 1970 John Horton Conway invented a new cellular automaton, which he prophetically named Life. In Life, cells exist in one of just two states: alive (black) or dead (white). The rules are astonishingly simple:

- A cell that is white at a given instant becomes black at the next if it has precisely three black neighbors.
- A cell that is black at a given instant becomes white at the next if it has four or more black neighbors.
- A cell that is black at a given instant becomes white at the next if it has one or no black neighbors.

—which are constructed with a view to denying them that power—you can say and do what you like; for they are then wholly impotent for mischief, and will not remember a few minutes hence the incident for which they may be at this moment threatening you with death, nor the promises which you may have found it necessary to make in order to pacify their fury.

On the whole we get on pretty smoothly in our domestic relations, except in the lower strata of the Military Classes. There the want of tact and discretion on the part of the husbands produces at times indescribable disasters. Relying too much on the offensive weapons of their acute angles instead of the defensive organs of good sense and seasonable simulations, these reckless creatures too often neglect the prescribed construction of the women's apartments, or irritate their wives by ill-advised expressions out of doors, which they refuse immediately to retract. Moreover a blunt and stolid regard for literal truth indisposes them to make those lavish promises by which the more judicious Circle can in a moment pacify his consort. The result is massacre; not, however, without its advantages, as it eliminates the more brutal and troublesome of the Isosceles; and by many of our Circles the destructiveness of the Thinner Sex is regarded as one among many providential arrangements for suppressing redundant population, and nipping Revolution in the bud.

Yet even in our best regulated and most approximately Circular families I cannot say that the ideal of family life is so high as with you in Spaceland. There is peace, in so far as the absence of slaughter may be called by that name, but there is necessarily little harmony of tastes or pursuits; and the cautious wisdom of the Circles has ensured safety at the cost of domestic comfort. In every Circular or Polygonal household it has been a habit from time immemorial—and now has become a kind of instinct among the women of our higher classes—that the mothers and daughters should constantly keep their eyes and mouths towards their husband and his male friends; and for a lady in a family of distinction to turn her back upon her husband would be regarded as a kind of portent, involving loss of *status*. But, as I shall soon shew, this custom, though it has the advantage of safety, is not without its disadvantages.

In the house of the Working Man or respectable Tradesman—where the wife is allowed to turn her back upon her husband, while pursuing her household avocations—there are at least intervals of quiet, when the wife is neither seen nor heard, except for the humming sound of the continuous Peace-cry; but in the homes of the upper classes there is too often no peace. There the voluble mouth and bright penetrating eye are ever directed towards the Master of the household; and light itself is not more persistent than the stream of feminine discourse. The tact and skill which suffice to avert a Woman's sting are unequal to the

The "Game of Life" starts with an object, or a cluster of objects, made from a configuration of black cells on a white-cell background. Then the automaton follows the rules, and you find out what happens. Many configurations die out completely (all cells white), become static, or cycle through a small number of states. Some configurations, such as the glider (Figure 18) *move*. Eventually Conway managed to prove that it is possible to use gliders to construct a Universal Turing Machine — an abstract programmable computer — from the rules of Life. Pulses of gliders carry "information" through the Turing machine.

Accordingly, if the citizens of Flatland had brains based on two-dimensional cellular automata, they could carry out any algorithmic computation. It is possible that our own minds are nonalgorithmic. This is a central contention of Roger Penrose's *The Emperor's New Mind,* but Penrose's arguments have been challenged from a variety of perspectives and certainly are not conclusive on this point. At any rate, Flatlanders with cellular-automaton brains could be as intelligent as the most accomplished artificial intelligence (AI) that the human race will ever build, which isn't bad for A. Square.

What of Flatland women? We now know that even a one-dimensional array of cells, communicating with their neighbors according to simple rules, can also carry out any task that a computer can accomplish, so actually there is no mathematical reason why the brains of Flatland females need be limited by their lower dimensionality.

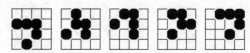

Figure 18 The glider, a mobile configuration of the Game of Life.

5 This paragraph concludes an extensive section of social satire, and the final sentence is particularly biting. Victorian society had a very clear idea of what was suitable for women, and intellectual activities, especially mathematics, in no way fitted those preconceptions. In the 1800s, Cambridge University refused to let women qualify for a degree. However, they were permitted to take the examinations on an informal basis. That is, their answers would be graded by the examiners, but only to indicate what level of understanding they had reached: They would not be given any formal credit. In the British university system of that period, the student was given only a single overall grade — first, second, or third class. Five women sat the 1870 examinations, and four of them got first-class marks, but their names did not appear in the "class lists," the list of grades, because they were not permitted to be official candidates for a degree. By 1872 one woman had passed the entire Mathematical Tripos, the full degree course, but still the official position was that such a performance did not merit the award of a degree if it was achieved by a woman. In 1881 Ms. C. A. Scott was placed level with the eighth wrangler (at Cambridge University, a wrangler is someone who obtains a first-class degree in mathematics, so the eighth wrangler is the eighth person in the pass list), and in 1890 Philippa Fawcett was placed above the senior wrangler, making her the best student of the entire class. Neither was awarded a degree. Only in 1920 were women finally permitted to take degrees at Cambridge, and another fifty years were to elapse before individual colleges admitted students of both sexes.

task of stopping a Woman's mouth; and as the wife has absolutely nothing to say, and absolutely no constraint of wit, sense, or conscience to prevent her from saying it, not a few cynics have been found to aver that they prefer the danger of the death-dealing but inaudible sting to the safe sonorousness of a Woman's other end.

To my readers in Spaceland the condition of our Women[5] may seem truly deplorable, and so indeed it is. A Male of the lowest type of the Isosceles may look forward to some improvement of his angle, and to the ultimate elevation of the whole of his degraded caste; but no Woman can entertain such hopes for her sex. "Once a Woman, always a Woman" is a Decree of Nature; and the very Laws of Evolution seem suspended in her disfavour. Yet at least we can admire the wise Pre-arrangement which has ordained that, as they have no hopes, so they shall have no memory to recall, and no forethought to anticipate, the miseries and humiliations which are at once a necessity of their existence and the basis of the constitution of Flatland.

§ 5. — *Of our Methods of Recognizing one another.*

You, who are blessed with shade as well as light, you, who are gifted with two eyes, endowed with a knowledge of perspective, and charmed with the enjoyment of various colours, you, who can actually see an angle, and contemplate the complete circumference of a Circle in the happy region of the Three Dimensions—how shall I make clear to you the extreme difficulty[1] which we in Flatland experience in recognizing one another's configuration?

Recall what I told you above. All beings in Flatland, animate or inanimate, no matter what their form, present *to our view* the same, or nearly the same, appearance, viz. that of a straight Line. How then can one be distinguished from another, where all appear the same?

The answer is threefold. The first means of recognition is the sense of hearing; which with us is far more highly developed than with you, and which enables us not only to distinguish by the voice our personal friends, but even to discriminate between different classes, at least so far as concerns the three lowest orders, the Equilateral,

1 Physics in a two-dimensional universe is subject to different constraints from those that impinge on a three-dimensional universe, and this affects the abilities of Flatlanders. Abbott discusses several examples of such constraints and makes a pretty good (though flawed) attempt to set up a consistent framework for two-dimensional lifeforms. The best sustained development of convincing two-dimensional physics, engineering, and biology is Dewdney's *The Planiverse.*

One flaw in Abbott's scheme, illustrating how physics varies with dimensionality, has been pointed out by the physicist Alan Solomon. The Dutch physicist Christian Huygens (1629–1695) showed that in three-dimensional space, sound waves move "sharply." If we are at some distance from a gunshot, we hear silence at first, then a sharply defined bang, then silence again. Solomon's point is that sound in two dimensions would be different. Again, we would at first hear silence, then a bang — but the bang would not be followed by silence. Instead, it would reverberate indefinitely, like a sustained echo that gradually faded. More generally, the mathematics of the "wave equation" shows that in *odd*-dimensional spaces, a sharp noise generated at a point propagates as a single sharp wavefront, whereas in *even*-dimensional spaces, a sharp noise generated at a point produces a system of propagating waves that reverberates forever. Flatland's dimension is even, so all Flatland sounds would reverberate. It would be like living inside a tin can.

the Square, and the Pentagon—for of the Isosceles I take no account. But as we ascend in the
social scale, the process of discriminating and being discriminated by hearing increases in difficulty, partly because voices are assimilated, partly
because the faculty of voice-discrimination is a
plebeian virtue not much developed among the
Aristocracy. And wherever there is any danger
of imposture we cannot trust to this method.
Amongst our lowest orders, the vocal organs are
developed to a degree more than correspondent
with those of hearing, so that an Isosceles can easily feign the voice of a Polygon, and, with some
training, that of a Circle himself. A second method
is therefore more commonly resorted to.

Feeling is, among our Women and lower classes
—about our upper classes I shall speak presently
—the principal test of recognition, at all events
between strangers, and when the question is, not
as to the individual, but as to the class. What
therefore "introduction" is among the higher classes
in Spaceland, that the process of "feeling" is with
us. "Permit me to ask you to feel and be felt by my
friend Mr. So-and-so"—is still, among the more
old-fashioned of our country gentlemen in districts remote from towns, the customary formula
for a Flatland introduction. But in the towns, and
among men of business, the words "be felt by" are
omitted and the sentence is abbreviated to, "Let
me ask you to feel Mr. So-and-so"; although it is
assumed, of course, that the "feeling" is to be re-

ciprocal. Among our still more modern and dashing young gentlemen—who are extremely averse to superfluous effort and supremely indifferent to the purity of their native language—the formula is still further curtailed by the use of "to feel" in a technical sense, meaning, "to recommend-for-the-purposes-of-feeling-and-being-felt"; and at this moment the "slang" of polite or fast society in the upper classes sanctions such a barbarism as "Mr. Smith, permit me to feel Mr. Jones."

Let not my Reader however suppose that "feeling" is with us the tedious process that it would be with you, or that we find it necessary to feel right round all the sides of every individual before we determine the class to which he belongs. Long practice and training, begun in the schools and continued in the experience of daily life, enable us to discriminate at once by the sense of touch, between the angles of an equal-sided Triangle, Square, and Pentagon;[2] and I need not say that the brainless vertex of an acute-angled Isosceles is obvious to the dullest touch. It is therefore not necessary, as a rule, to do more than feel a single angle of an individual; and this, once ascertained, tells us the class of the person whom we are addressing, unless indeed he belongs to the higher sections of the nobility. There the difficulty is much greater. Even a Master of Arts in our University of Wentbridge has been known to confuse a ten-sided with a twelve-sided Polygon; and there is hardly a Doctor of Science in or out of that fa-

2 In this passage Abbott refers to polygons with 3, 4, 5, 10, 12, 20, and 24 sides; elsewhere the hexagon, with 6 sides, is also mentioned. Only much later in the book are polygons with 300, 400, 600, and 10,000 sides referred to, and then only in passing as examples of "lots of sides." Why do no polygons with 7 or 9 sides appear, for example? After all, by Abbott's own statement, the son of a Hexagon should be a Heptagon: There ought to be plenty of them around. The numerology may be coincidental, but Abbott was a schoolmaster, and his choice of polygons would be influenced by the ones that are commonly encountered in school geometry. In Euclid, emphasis is placed on the construction of figures using the traditional instruments of ruler and compasses, although later Greek geometers permitted a wider range of instruments. Euclid's *Elements* include (ruler-and-compass) constructions for the equilateral triangle ([1,1]; Figure 19a), square (see Figure 1 in Book [1,46]), and pentagon ([4,11]; Figure 19b). There is also a construction for bisecting an angle (dividing it into two equal halves; Book I, Proposition 9). By combining these constructions, regular polygons can be constructed with 3, 4, 5, 6, 8, 10, 12, 15, 16, 20, 24, 30 … sides—the numbers 1, 3, 5, and 15 multiplied by any power of 2 (but excluding polygons with 1 and 2 sides because these reduce to a point and a line segment). All of the

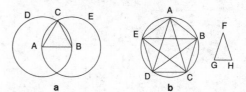

Figure 19 Euclid's constructions for an equilateral triangle and a regular pentagon. (**a**) Given a line AB, draw a circle BCD with center A and radius AB. Then draw a circle ACE with center B and radius AB. The circles meet at C, and ABC is equilateral. (**b**) Construct an isosceles triangle FGH with angle FGH twice angle GFH (Euclid has already explained how to do this in [4, 10]). Inscribe triangle ACD similar to FGH. Bisect angle ACD to obtain E, and similarly obtain B. Then ABCDE is a regular pentagon.

polygons Abbott mentions here are *constructible,* whereas the 300-, 400-, 600-, and 10,000-sided polygons mentioned later are not. His later discussion of parents performing surgery on their children to increase the number of sides seems to indicate that the sides are all bisected, again a feasible Euclidean construction (*Elements* [1,10]). Thus it looks as though Abbott lifted some of the Flatland sociology straight out of Euclid.

Euclid does not include ruler-and-compass constructions for regular polygons with 7, 9, 11, 13 … sides, and for roughly 2,000 years it was believed that the only constructible regular polygons were those known to the Greeks. In 1796 the young Gauss, then age 19, was trying to decide whether to study mathematics or philology (a branch of linguistics that deals with the historical development

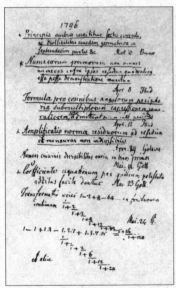

Figure 20 Gauss's notebook. The first entry reads: "Principia quibus innititus sectio arculi, ac divisibilitas eiusdem geometrica in Septemdecim partes &c." ("Principles by which to obtain sections of an arc, such as the geometric divisibility of the same into seventeen parts, etc.")

mous University who could pretend to decide promptly and unhesitatingly between a twenty-sided and a twenty-four sided member of the Aristocracy.

Those of my readers who recall the extracts I gave above from the Legislative code concerning Women, will readily perceive that the process of introduction by contact requires some care and discretion. Otherwise the angles might inflict on the unwary Feeler irreparable injury. It is essential for the safety of the Feeler that the Felt should stand perfectly still. A start, a fidgety shifting of the position, yes, even a violent sneeze, has been known before now to prove fatal to the incautious, and to nip in the bud many a promising friendship. Especially is this true among the lower classes of the Triangles. With them, the eye is situated so far from their vertex that they can scarcely take cognizance of what goes on at that extremity of their frame. They are, moreover, of a rough coarse nature, not sensitive to the delicate touch of the highly organized Polygon. What wonder then if an involuntary toss of the head has ere now deprived the State of a valuable life!

I have heard that my excellent Grandfather—one of the least irregular of his unhappy Isosceles class, who indeed obtained, shortly before his decease, four out of seven votes from the Sanitary and Social Board for passing him into the class of the Equal-sided—often deplored, with a tear in his venerable eye, a miscarriage of this kind, which had occurred to his great-great-great-Grandfather, a respectable Working Man with an angle or brain of 59°30'.[3] According to his account, my unfortunate Ancestor, being afflicted with rheumatism, and in the act of being felt by a Polygon, by one sudden start accidentally transfixed the Great Man through the diagonal; and thereby, partly in consequence of his long imprisonment and degradation, and partly because of the moral shock which pervaded the whole of my Ancestor's relations, threw back our family a degree and a half in their ascent towards better things. The result was that in the next generation the family brain was registered at only 58°, and not till the lapse of five generations was the lost ground recovered, the full 60° attained, and the Ascent from the Isosceles finally achieved. And all this series of calamities from one little accident in the process of Feeling.

At this point I think I hear some of my better educated readers exclaim, "How could you in Flatland know anything about angles and degrees, or minutes? We can *see* an angle, because we, in the region of Space, can see two straight lines inclined to one another; but you, who can see nothing but one straight line at a time, or at all events only a number of bits of straight lines all in one straight line, — how can you ever discern any angle, and much less register angles of different sizes?"

I answer that though we cannot *see* angles,[4] we can *infer* them, and this with great precision. Our

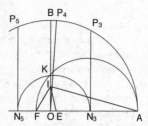

Figure 21 A ruler-and-compass construction for the regular 17-gon. Let OA and OB be perpendicular radii of a circle. Let IO be one-quarter of the way from O to B, and make angle OIE one-quarter of angle OIA. Make angle EIF equal to 45°. The circle with diameter AF cuts OB at K. The circle with center E that passes through K cuts AO at N_3 and N_5. Draw N_3P_3 and N_5P_5 perpendicular to OA. Then P_3 and P_5 are the third and fifth vertices of a regular 17-gon. Bisect angle P_3OP_5 to locate P_4. Then mark off successive edges at distances P_4P_5 along the circle to locate all other vertices.

of words). On March 30 of that year, his mind was made up by an epic discovery: There exists a ruler-and-compass construction for the regular 17-sided polygon. He started a mathematical notebook, and this was the first entry (Figure 20). Later, an explicit construction was worked out by following Gauss's algebraic recipe (Figure 21).

More generally, Gauss proved that such constructions exist if and only if the number of sides of the polygon is a power of 2 multiplied by (zero or more) distinct *Fermat primes.* These primes are named after the French mathematician Pierre de Fermat (1601–1665); their defining characteristic is that they are 1 greater than a power of 2. Some such numbers are prime, others are not, and Fermat looked for a pattern to distinguish the primes. He quickly realized that the only possible candidates are the numbers $2 + 1$, $2^2 + 1$, $2^4 + 1$, $2^8 + 1$, $2^{16} + 1$, and so on, where the powers that occur are themselves powers of 2. These are the Fermat numbers; any that happen to be prime are called Fermat primes. In several letters from 1640 onward, Fermat pointed out that the first five Fermat numbers, 3, 5, 17, 257, and 65,537, are all prime. He conjectured that every Fermat number is prime, but in 1732 the Swiss-born mathematician Leonhard Euler (1707–1783) pointed out that the sixth Fermat number, 4,294,967,297, factorizes as 641 x 6,700,417 and so is not prime. Indeed, no other Fermat primes are known, but some gigantic Fermat prime may yet be discovered.

Gauss's work implies that the constructible regular polygons with fewer than 100 sides have the following numbers of sides: 3, 4, 5, 6, 8, 10, 12, 15, 16, 17, 20, 24, 30, 32, 34, 40, 48, 51, 60, 64, 68, 80, 85, and 96. An explicit construction for the regular 257-gon was found by F. J. Richelot in 1832. It is said that a certain Prof. Hermes of Lingen spent ten years working on a construction for the regular 65,537-gon, and his papers are deposited in the University of Göttingen. Perhaps the most important consequence of all this for mathematics was that Gauss's discovery of the constructibility of the regular 17-gon convinced him to become a mathematician rather than a philologist. He went on to become the greatest mathematician of all time.

3 Here ° represents "degree" and ' is "minute" — one-sixtieth of a degree — so 30' is half a degree. The angles of an equilateral triangle are 60°, one sixth of a full circle of 360°. This follows because the angles of any triangle add up to 180° (*Elements* [1,32]) and all three angles are equal. A. Square's excellent Grandfather's great-great-great-Grandfather had almost attained Regularity, being only half a degree short of the regular 60° angle. The family could reasonably have expected the next generation to reach that desirable goal (a gain of half a degree per generation is specified, and this figure is consistent with the numbers in this passage), but the accident set the family back to 58° instead. It therefore took a further four generations, making five in all, to attain Equilateralhood.

The convention that a full circle amounts to 360° goes back to the Babylonians, whose number system was "base 60," whereas ours is "base 10." That is, the values of symbols in the "units" position are multiplied by 1, those in the next "60s" position — to the left — are multiplied by 60, and so on. Babylon-

sense of touch, stimulated by necessity, and developed by long training, enables us to distinguish angles far more accurately than your sense of sight, when unaided by a rule or measure of angles. Nor must I omit to explain that we have great natural helps. It is with us a Law of Nature that the brain of the Isosceles class shall begin at half a degree, or thirty minutes, and shall increase (if it increases at all) by half a degree in every generation,[5] until the goal of 60° is reached, when the condition of serfdom is quitted, and the freeman enters the class of Regulars.

Consequently, Nature herself supplies us with an ascending scale or Alphabet of angles for half a degree up to 60°, Specimens of which are placed in every Elementary School throughout the land. Owing to occasional retrogressions, to still more frequent moral and intellectual stagnation, and to the extraordinary fecundity of the Criminal and Vagabond Classes, there is always a vast superfluity of individuals of the half degree and single degree class, and a fair abundance of Specimens up to 10°. These are absolutely destitute of civic rights; and a great number of them, not having even intelligence enough for the purposes of warfare, are devoted by the States to the service of education. Fettered immovably so as to remove all possibility of danger, they are placed in the class rooms of our Infant Schools, and there they are utilized by the Board of Education[6] for the purpose of imparting to the offspring of the Middle

Classes that tact and intelligence of which these wretched creatures themselves are utterly devoid.

In some States the Specimens are occasionally fed and suffered to exist for several years; but in the more temperate and better regulated regions, it is found in the long run more advantageous for the educational interests of the young, to dispense with food, and to renew the Specimens every month—which is about the average duration of the foodless existence of the Criminal class. In the cheaper schools, what is gained by the longer existence of the Specimen is lost, partly in the expenditure for food, and partly in the diminished accuracy of the angles, which are impaired after a few weeks of constant "feeling." Nor must we forget to add, in enumerating the advantages of the more expensive system, that it tends, though slightly yet perceptibly, to the diminution of the redundant Isosceles population—an object which every statesman in Flatland constantly keeps in view. On the whole therefore—although I am not ignorant that, in many popularly elected School Boards, there is a reaction in favour of "the cheap system" as it is called—I am myself disposed to think that this is one of the many cases in which expense is the truest economy.

But I must not allow questions of School Board politics to divert me from my subject. Enough has been said, I trust, to shew that Recognition by Feeling is not so tedious or indecisive a process as might have been supposed; and it is obviously

ian writing was cuneiform—composed of wedge-shaped impressions in clay made with a stylus that had a triangular cross section—and in the symbolism of the Akkadian period, the inscription ⟨symbol⟩ represents $1 . 60^2 + 0.60 + 3$—that is, 3,603. The special symbol ⟨symbol⟩ is a separator, indicating no entry in the 60s place; this is the first known occurrence of a concept of "zero" in mathematics. The positional notation extended to fractions, with successive places to the right of the units position indicating multiples of $1/60$, $1/3600$, and so on.

Around 2400 B.C. the Sumerians, who lived in the same part of the world as the Babylonians, had used a year of $30 \times 12 = 360$ days. In the late Babylonian period, the first century B.C., the division of the circle into $6 \times 60 = 360$ units was developed, probably because 360 is roughly the number of days in a year. Today's convention that a minute is $1/60$ of an hour and a second is $1/60$ of a minute goes back to the Babylonians, as does the division of a degree into minutes and seconds. (The terminology is instructive: A minute is a *small* part, and a second is a *second* subdivision of the basic unit. This is reflected in the notation: 5' means "5 minutes" and 5" means "five seconds." If we ever got to "thirds," as the Babylonians did, we'd write them like this: 5'''.) For theoretical purposes, mathematicians prefer the *radian* as a unit of angular measurement. A radian is the angle corresponding to an arc of unit length in a circle of unit radius, and it is roughly 57° 17' (and exactly $180/\pi$ degrees). Various basic formulas of the calculus are much simpler when radian measurement is used.

4 We Spacelanders experience the analogous problem: obtaining accurate perceptions of a three-dimensional world from its projection onto a two-dimensional retina (more precisely, two stereoscopically related projections onto two retinas). The human visual system works so well that we obtain a vivid illusion of a thoroughly three-dimensional world from astonishingly limited sense organs. For example, the part of our field of vision in which the image appears "sharp" is about the width of the word sharp when this book is held in normal reading position. (Stare at the word and observe that neighboring words appear blurred.) Science has only just begun to unravel the marvelous trickery that the brain uses to analyze the visual field and create a vivid neural model of reality.

5 This supports the calculations in note [3] of chapter 5. Without the parenthetic statement "if it increases at all," the brain (angle) of the Isosceles class begins at half a degree, so after 720 generations all Isosceles triangles would have become Equilateral, and after another few thousand generations there would be nothing but Circles, which you will recall in Flatland means polygons with a huge number of very tiny sides. What does *begins* mean here, though? Can new members of the Isosceles class arise spontaneously? Or was Flatland *created* (which would appeal to Abbott's theological beliefs) —possibly with one very thin Isosceles and one Line? This would be a "Garden of Eden configuration" in the mathematical terminology of cellular automata, which means that it has no precursor according to the dynamical rules. Because of these difficulties, and for other narrative reasons, I was forced to assert in *Flatterland* that A. Square's "invariable law" of increase in regularity was an exaggeration.

6 Abbott is clearly protesting some Victorian educational practice … but what? Inclusion of the educationally subnormal in ordinary classes? Tokenism? Such practices certainly existed in his day. For example, in 1825 the Common Council of the Corporation of London recommended that "Carpenter's Children" — the boys being educated (at Tonbridge School) through the bequest of John Carpenter — should wear a special uniform of "blue Jacket and Trowsers," and that "a small silver Medal should always be worn by the said Boys denoting the objects of the said Charity." This insensitive suggestion was vetoed by Dr. Knox, the master of Tonbridge School.

more trustworthy than Recognition by hearing. Still there remains, as has been pointed out above, the objection that this method is not without danger. For this reason many in the Middle and Lower classes, and all without exception in the Polygonal and Circular orders, prefer a third method, the description of which shall be reserved for the next section.

§ 6.—Of Recognition by Sight.

I AM ABOUT to appear very inconsistent. In previous sections I have said that all figures in Flatland present the appearance of a straight line; and it was added or implied, that it is consequently impossible to distinguish by the visual organ between individuals of different classes: yet now I am about to explain to my Spaceland critics how we are able to recognize one another by the sense of sight.

If however the Reader will take the trouble to refer to the passage in which Recognition by Feeling is stated to be universal, he will find this qualification—"among the lower classes." It is only among the higher classes and in our more temperate climates that Sight Recognition is practised.

That this power exists in any regions and for any classes is the result of Fog;[1] which prevails during the greater part of the year in all parts save the torrid zones. That which is with you in Spaceland an unmixed evil, blotting out the landscape, depressing the spirits, and enfeebling the health, is by us recognized as a blessing scarcely inferior to air itself, and as the Nurse of arts and Parent

1 Flatlanders perceive a two-dimensional world through a one-dimensional projection; similarly, humans perceive a three-dimensional world through a two-dimensional projection. The big difficulty in both cases is *depth perception*—reconstructing the "missing information" along the direction of projection. Humans achieve this using a variety of cues: stereoscopic vision (each eye receives a slightly different image, related by depth), apparent size of objects (distant objects look smaller because of perspective), and overlaps (if one object overlaps another, it is closer). All of these tricks have two-dimensional analogues, but Abbott chooses to introduce a different solution: fog. Fog causes distant objects to look fainter. This is one of his least satisfactory inventions because the fog he describes must be fairly dense to offer effective depth perception at short range. This means that anything distant will be totally obscured by fog, yet elsewhere he describes scenes (such as the impressive maneuvers of ranks of soldiers) that cannot be so obscured. Having eliminated the problem of shadows by appealing to an external light source, he accidentally reintroduces a similar problem by resorting to fog. We are reminded that *Flatland* is a fantasy, not a logically consistent scientific treatise.

2 Abbott's approach to visual perception is very static: The Flatlander's eye observes a single snapshot of the approaching object, from a single direction. Rotations, or other movement, would supply additional cues, just as they do in Spaceland.

of sciences. But let me explain my meaning, without further eulogies on this beneficent Element.

If Fog were non-existent, all lines would appear equally and indistinguishably clear; and this is actually the case in those unhappy countries in which the atmosphere is perfectly dry and transparent. But wherever there is a rich supply of Fog, objects that are at a distance, say of three feet, are appreciably dimmer than those at a distance of two feet eleven inches; and the result is that by careful and constant experimental observation of comparative dimness and clearness, we are enabled to infer with great exactness the configuration of the object observed.

An instance will do more than a volume of generalities to make my meaning clear.

Suppose I see two individuals approaching[2] whose rank I wish to ascertain. They are, we will suppose, a Merchant and a Physician, or in other words, an Equilateral Triangle and a Pentagon: how am I to distinguish them?

It will be obvious, to every child in Spaceland who has touched the threshold of Geometrical Studies, that, if I can bring my eye so that its glance may bisect an angle (A) of the approaching stranger, my view will lie as it were evenly between his two sides that are next to me (viz. CA and AB), so that I shall contemplate the two impartially, and both will appear of the same size.

Now in the case of (1) the Merchant, what shall I see? I shall see a straight line DAE, in which the

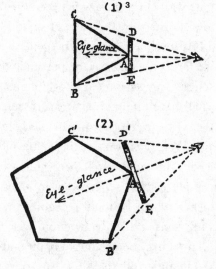

3 In the first edition caption (1) is missing, and so is the letter A at the vertex of the Triangle.

4 In the first edition there is an error; it has AD instead of AB.

middle point (A) will be very bright because it is nearest to me; but on either side the line will shade away *rapidly into dimness,* because the sides AC and AB *recede rapidly*[4] *into the fog* and what appear to me as the Merchant's extremities, viz. D and E, will be *very dim indeed.*

On the other hand in the case of (2) the Physician, though I shall here also see a line (D'A'E') with a bright centre (A'), yet it will shade away *less rapidly* into dimness, because the sides (A'C', A'B') *recede less rapidly into the fog*: and what appear to me the Physician's extremities, viz. D' and E', will not be *not so dim* as the extremities of the Merchant.

The Reader will probably understand from these two instances how — after a very long training supplemented by constant experience — it is possible for the well-educated classes among us to discriminate with fair accuracy between the middle and lowest orders, by the sense of sight. If my Spaceland Patrons have grasped this general conception, so far as to conceive the possibility of it and not to reject my account as altogether incredible—I shall have attained all I can reasonably expect. Were I to attempt further details I should only perplex. Yet for the sake of the young and inexperienced, who may perchance infer—from the two simple instances I have given above, of the manner in which I should recognize my Father and my Sons—that Recognition by sight is an easy affair, it may be needful to point out that in actual life most of the problems of Sight Recognition are far more subtle and complex.

If for example, when my Father, the Triangle, approaches me, he happens to present his side to me instead of his angle, then, until I have asked him to rotate, or until I have edged my eye round him, I am for the moment doubtful whether he may not be a Straight Line, or, in other words, a Woman. Again, when I am in the company of one of my two hexagonal Grandsons, contemplating one of his sides (AB) full front, it will be evident from the accompanying diagram that I shall

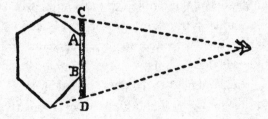

see one whole line (AB) in comparative brightness (shading off hardly at all at the ends) and two smaller lines (CA and BD) dim throughout and shading away into greater dimness towards the extremities C and D.

But I must not give way to the temptation of enlarging on these topics. The meanest mathematician in Spaceland will readily believe me when I assert that the problems of life, which present themselves to the well-educated—when they are themselves in motion, rotating, advancing or retreating, and at the same time attempting to discriminate by the sense of sight between a number of Polygons of high rank moving in different directions, as for example in a ball-room or conversazione—must be of a nature to task the angularity of the most intellectual, and amply justify the rich endowments of the Learned Professors of Geometry,[5] both Static and Kinetic, in the illustrious University of Wentbridge, where the Science and Art of Sight Recognition are regularly taught to large classes of the *élite* of the States.

It is only a few of the scions of our noblest and wealthiest houses, who are able to give the time and money necessary for the thorough prosecution of this noble and valuable Art. Even to me, a Mathematician of no mean standing, and the Grandfather of two most hopeful and perfectly regular Hexagons, to find myself in the midst of a crowd of rotating Polygons of the higher classes,

5 The bridge is manifestly Cambridge, Abbott's alma mater. He was a student at St. John's College, Cambridge, from 1857 to 1861, and in 1862 he became a college fellow. The Went is a pun: *Came*-bridge/*Went*-bridge. There is a river Went in Yorkshire but that is probably a coincidence.

is occasionally very perplexing. And of course to a common Tradesman, or Serf, such a sight is almost as unintelligible as it would be to you, my Reader, were you suddenly transported into our country.

In such a crowd you could see on all sides of you nothing but a Line, apparently straight, but of which the parts would vary irregularly and perpetually in brightness or dimness. Even if you had completed your third year in the Pentagonal and Hexagonal classes in the University, and were perfect in the theory of the subject, you would still find that there was need of many years of experience, before you could move in a fashionable crowd without jostling against your betters, whom it is against etiquette to ask to "feel," and who, by their superior culture and breeding, know all about your movements, while you know very little or nothing about theirs. In a word, to comport oneself with perfect propriety in Polygonal society, one ought to be a Polygon oneself. Such at least is the painful teaching of my experience.

It is astonishing how much the Art—or I may almost call it instinct—of Sight Recognition is developed by the habitual practice of it and by the avoidance of the custom of "Feeling." Just as, with you, the deaf and dumb, if once allowed to gesticulate and to use the hand-alphabet, will never acquire the more difficult but far more valuable art of lip-speech and lip-reading, so it is with us as regards "Seeing" and "Feeling." None who in early life resort to "Feeling" will ever learn "Seeing" in perfection.

For this reason, among our Higher Classes, "Feeling" is discouraged or absolutely forbidden. From the cradle their children, instead of going to the Public Elementary schools (where the art of Feeling is taught,) are sent to higher Seminaries of an exclusive character; and at our illustrious University, to "feel" is regarded as a most serious fault, involving Rustication for the first offence, and Expulsion for the second.

But among the lower classes the art of Sight Recognition is regarded as an unattainable luxury. A common Tradesman cannot afford to let his son spend a third of his life in abstract studies. The children of the poor are therefore allowed to "feel" from their earliest years, and they gain thereby a precocity and an early vivacity which contrast at first most favourably with the inert, undeveloped, and listless behaviour of the half-instructed youths of the Polygonal class; but when the latter have at last completed their University course, and are prepared to put their theory into practice, the change that comes over them may almost be described as a new birth, and in every art, science, and social pursuit they rapidly overtake and distance their Triangular competitors.

Only a few of the Polygonal Class fail to pass the Final Test or Leaving Examination at the University. The condition of the unsuccessful minority is truly pitiable. Rejected from the higher class,

6 The view of heredity here is decidedly Lamarckian: Failure to pass the Leaving Examination at the University translates into irregular offspring. Abbott is probably opposing the view, prevalent at his time, that social status (in the class hierarchy) is more important than innate ability. For instance, according to A. G. Howson's *A History of Mathematics Education in England,*

> Neither was the cause [of professional training of teachers] helped by such pronouncements as [E.E.] Bowen's (Headmaster of Harrow) to the 1895 Bryce Commission: "A bad man teaching history well is a far worse thing than a good man teaching history badly."

By *good man,* Bowen meant a man of "character" — as determined by the standards and prejudices of the upper classes of his day.

7 The image here is of failures as a source of sedition. The association of college dropouts with non-conformity was alive and well in Victorian times.

they are also despised by the lower. They have neither the matured and systematically trained powers of the Polygonal Bachelors and Masters of Arts, nor yet the native precocity and mercurial versatility of the youthful Tradesman. The professions, the public services, are closed against them; and though in most States they are not actually debarred from marriage, yet they have the greatest difficulty in forming suitable alliances, as experience shews that the offspring of such unfortunate and ill-endowed parents is generally itself unfortunate, if not positively Irregular.[6]

It is from these specimens of the refuse of our Nobility[7] that the great Tumults and Seditions of past ages have generally derived their leaders; and so great is the mischief thence arising that an increasing minority of our more progressive Statesmen are of opinion that true mercy would dictate their entire suppression, by enacting that all who fail to pass the Final Examination of the University should be either imprisoned for life, or extinguished by a painless death.

But I find myself digressing into the subject of Irregularities, a matter of such vital interest that it demands a separate section.

§ 7. — *Concerning Irregular Figures.*[1]

THROUGHOUT the previous pages I have been assuming[2]—what perhaps should have been laid down at the beginning as a distinct and fundamental proposition—that every human being in Flatland is a Regular Figure, that is to say of regular construction. By this I mean that a Woman must not only be a line, but a straight line; that an Artisan or Soldier must have two of his sides equal; that Tradesmen must have three sides equal; Lawyers (of which class I am a humble member), four sides equal, and, generally, that in every Polygon, all the sides must be equal.

The size of the sides would of course depend upon the age of the individual. A Female at birth would be about an inch long, while a tall adult Woman might extend to a foot. As to the Males of every class, it may be roughly said that the length of an adult's sides, when added together, is two feet or a little more. But the size of our sides is not under consideration. I am speaking of the *equality* of sides, and it does not need much reflection to see that the whole of the social life in Flatland rests upon the fundamental fact that Nature wills all Figures to have their sides equal.

1 Here the first edition has "*§7.—Of Irregular Figures.*" However, the contents list in both editions has "*7 Concerning Irregular Figures.*"

2 The deep mathematical concept underlying regularity of polygons is that of *symmetry*. A symmetry of a geometric shape is a transformation (here a rigid motion, such as a translation, rotation, or reflection) that leaves the shape unchanged.

Every shape, however irregular, has at least one symmetry: the transformation "do nothing." The isosceles triangle has a second symmetry: "reflect about a line at right angles to the base." An equilateral triangle has six symmetries: three rotations (by 0°, 120°, and 240°) and three reflections (in the lines that bisect the angles); see Figure 22a. In general, a regular n-gon has $2n$ symmetries, and in particular a square has eight; see Figure 22b. Thus the more symmetries a figure has, the higher its place in the Flatland hierarchy. A true Circle, with *infinitely many symmetries* (all rotations about the center and all reflections in diameters) must necessarily be the most perfect being of all. In this view of perfection, the Flatlanders share the prejudices of the ancient Greek geometers, to whom the circle was the most perfect two-dimensional form and the sphere was the most perfect three-dimensional form. Abbott's appeal to Regularity is really one to Symmetry.

The human mind seems to have an innate attraction to symmetric images, and this may be the result of an evolutionary phenomenon that intrigued Darwin:

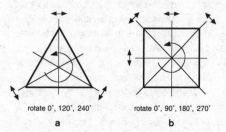

rotate 0°, 120°, 240°   rotate 0°, 90°, 180°, 270°

   a   **b**

Figure 22 (a) The six symmetries of an equilateral triangle. **(b)** The eight symmetries of a square.

sexual selection. The classic example is the peacock's ornate tail. Darwin's theory was that by chance, a few peacocks developed slightly larger and more decorative tails than their ancestors had had. If at much the same time, a few peahens also happened to develop an innate preference for ornate tails, then males with ornate tails would have a greater chance of having offspring, and so would females who *preferred* males with ornate tails. In *On The Origin of Species,* he writes,

> this leads me to say a few words on what I call Sexual Selection. This depends, not on a struggle for existence, but on a struggle between the males for possession of the females; the result is not death to the unsuccessful competitor, but few or no offspring.... The rock-thrush of Guiana, birds of Paradise, and some others, congregate; and successive males display their gorgeous plumage and perform strange antics before the females, which standing by as spectators, at last choose the most attractive partner.... I can see no good reason to doubt that female birds, by selecting during thousands of generations, the most melodious or beautiful males, according to their standard of beauty, might produce a marked effect.

At this point what Richard Dawkins later called an evolutionary arms race would set in: a runaway feedback loop leading to ever-more-ornate tails. Amotz Zahavi added a further observation, partly presaged by Darwin: It is no coincidence that large, ornate tails are preferred rather than small, dull ones. To a bird, an over-large tail is a handicap, and a male must be genetically well endowed to be able to operate despite the handicap. Thus females who mate with big-tailed birds automatically pass on "good genes" to their offspring. The same is *not* the case for small tails. Note that neither males nor females need be aware that good genes go with big tails; "blind" natural selection is what makes the process work.

Another test for good genes is symmetry: An organism's development must function very well in order to make its left and right sides more or less

If our sides were unequal our angles might be unequal.[3] Instead of its being sufficient to feel, or estimate by sight, a single angle in order to determine the form of an individual, it would be necessary to ascertain each angle by the experiment of Feeling. But life would be too short for such a tedious groping. The whole science and art of Sight Recognition would at once perish; Feeling, so far as it is an art, would not long survive; intercourse would become perilous or impossible; there would be an end to all confidence, all forethought; no one would be safe in making the most simple social arrangements; in a word, civilization would relapse into barbarism.

Am I going too fast to carry my Readers with me to these obvious conclusions? Surely a moment's reflection, and a single instance from common life, must convince every one that our whole social system is based upon Regularity, or Equality of Angles. You meet, for example, two or three Tradesmen in the street, whom you recognize at once to be Tradesmen by a glance at their angles and rapidly bedimmed sides, and you ask them to step into your house to lunch. This you do at present with perfect confidence, because everyone knows to an inch or two the area occupied by an adult Triangle: but imagine that your Tradesman drags behind his regular and respectable vertex, a parallelogram of twelve or thirteen inches in diagonal: — what are you to do with such a monster sticking fast in your house door?

But I am insulting the intelligence of my Readers by accumulating details which must be patent to everyone who enjoys the advantages of a Residence in Spaceland. Obviously the measurements of a single angle would no longer be sufficient under such portentous circumstances; one's whole life would be taken up in feeling or surveying the perimeter of one's acquaintances. Already the difficulties of avoiding a collision in a crowd are enough to tax the sagacity of even a well-educated Square; but if no one could calculate the Regularity of a single figure in the company, all would be chaos and confusion, and the slightest panic would cause serious injuries, or — if there happened to be any Women or Soldiers present — perhaps considerable loss of life.

Expediency therefore concurs with Nature[4] in stamping the seal of its approval upon Regularity of conformation: nor has the Law been backward in seconding their efforts. "Irregularity of Figure" means with us the same as, or more than, a combination of moral obliquity and criminality with you, and is treated accordingly. There are not wanting, it is true, some promulgators of paradoxes who maintain that there is no necessary connection between geometrical and moral Irregularity.[5] "The Irregular," they say, "is from his birth scouted by his own parents, derided by his brothers and sisters, neglected by the domestics, scorned and suspected by society, and excluded from all posts of responsibility, trust, and

identical. Thus sexual selection for symmetry can also operate, and it is known that a preference for symmetrical tails has evolved in some birds. In humans, there is an established preference for symmetric faces (film stars' faces are unusually symmetric, for example) and probably for symmetric bodies too. This preference seems to have become generalized into an appreciation of "beauty" and a link between beauty and symmetry.

Ironically, a Flatland woman — which here we take to be a line segment — has *four* symmetries (considered as a subset of the plane, not the line):

- Do nothing.
- Rotate through 180°.
- Reflect in a mirror coincident with the line segment.
- Reflect in a mirror coincident with the line segment and rotate through 180°.

The first two symmetries have the same effect on the line segment but different effects on the surrounding plane, and the second two behave similarly. This is consistent with the interpretation of a line segment as a 2-gon: a polygon with two vertices (the ends of the line segment). Conventionally, a polygon is a *closed* curve, and to achieve this it is necessary to consider the segment as being described twice, once in each direction. In the hierarchy of Flatland, a woman thus lies between isosceles triangles (with two symmetries) and equilateral triangles (with six). Clearly this argument had not occurred to the male-dominated Flatland hierarchy, despite the prominence of mathematics in its society.

3 Abbott's attention to mathematical detail is once again in evidence in the words *might be.* Polygons with unequal sides do not *necessarily* have unequal angles. The rectangle is a simple example of one that does not, and the *parhexagon,* a hexagon with not-necessarily-equal parallel sides, can (but does not always) have all angles equal to 60° (Figure 23a). Having based social recognition in Flatland on the size of a person's *angles,* Abbott needs a

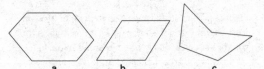

Figure 23 (**a**) A parhexagon. (**b**) A rhombus. (**c**) An equal-sided pentagon with unequal angles.

reason for all angles of any given individual to be equal. However, equality of sides is not a conclusive choice either because a polygon with all sides equal may have unequal angles. The rhombus (Figure 23b) is a four-sided figure with unequal angles, and from pentagons upward, many more such cases can arise (Figure 23c).

4 Here begins a lengthy passage satirizing beauty as a criterion for social desirability. However, the evolutionary link between beauty and "good genes" caused by sexual selection means that beauty is not so arbitrary a criterion as might be thought. It is, however, a criterion on the collective level: No particular individual's social worth can be inferred from that individual's physical beauty.

5 This is the long-running nature/nurture debate concerning whether such attributes as intelligence (and here morality) are governed by heredity or up-bringing — in modern terms, by genes or environment. Modern debates about such matters have acknowledged that the answer may be neither heredity nor environment alone, but some combination. However, such debates are still riddled with simplistic linear thinking — for example, scientists may claim that intelligence is 37% genetic and 63% environmental. Such thinking assumes that the effects of two distinct factors must be additive, whereas in practice, many more subtle interactions come into play. (It recently emerged that even within the linear paradigm, a third important factor had been totally overlooked: the condition of the mother's uterus. This makes all previous work in the area moot.) Lin-

useful activity. His every movement is jealously watched by the police till he comes of age and presents himself for inspection; then he is either destroyed, if he is found to exceed the fixed margin of deviation, or else immured in a Government Office as a clerk of the seventh class; prevented from marriage; forced to drudge at an uninteresting occupation for a miserable stipend; obliged to live and board at the office, and to take even his vacation under close supervision; what wonder that human nature, even in the best and purest, is embittered and perverted by such surroundings!"

All this very plausible reasoning does not convince me, as it has not convinced the wisest of our Statesmen, that our ancestors erred in laying it down as an axiom of policy that the toleration of Irregularity is incompatible with the safety of the State. Doubtless, the life of an Irregular is hard; but the interests of the Greater Number require that it shall be hard. If a man with a triangular front and a polygonal back were allowed to exist and to propagate a still more Irregular posterity, what would become of the arts of life? Are the houses and doors and churches in Flatland to be altered in order to accommodate such monsters? Are our ticket-collectors to be required to measure every man's perimeter before they allow him to enter a theatre, or to take his place in a lecture room? Is an Irregular to be exempted from the militia? And if not, how is he to be prevented from carrying desolation into the ranks of his comrades? Again, what irresistible temptations to fraudulent impostures must needs beset such a creature! How easy for him to enter a shop with his polygonal front foremost, and to order goods to any extent from a confiding tradesman! Let the advocates of a falsely called Philanthropy plead as they may for the abrogation of the Irregular Penal Laws, I for my part have never known an Irregular[6] who was not also what Nature evidently intended him to be — a hypocrite, a misanthropist, and, up to the limits of his power, a perpetrator of all manner of mischief.

Not that I should be disposed to recommend (at present) the extreme measures adopted in some States, where an infant whose angle deviates by half a degree from the correct angularity is summarily destroyed at birth. Some of our highest and ablest men, men of real genius, have during their earliest days laboured under deviations as great as, or even greater than, forty-five minutes: and the loss of their precious lives would have been an irreparable injury to the State. The art of healing also has achieved some of its most glorious triumphs in the compressions, extensions, trepannings, colligations, and other surgical or diaetetic operations by which Irregularity has been partly or wholly cured. Advocating therefore a *Via Media*, I would lay down no fixed or absolute line of demarcation; but at the period

ear thinking in this regard is reinforced by the existence of statistical techniques, known as analysis of variance (ANOVA), for resolving complex behavior into a sum of distinct factors, each weighted by a percentage. Practitioners of such techniques sometimes appear to be unaware that although they always give answers, their methodology is valid only when the component factors actually are additive. If you ask ANOVA software how this year's rainfall in Rome is related to the number of tigers in India during the nineteenth century and to the price of movie tickets in 1930s New York, and feed in the data, it will give an answer such as "Roman rainfall is 37.229% tigers and 62.771% movie tickets."

6 Banchoff says,

In the rigid society of Victorian England, there was little tolerance for irregularity. It was often associated with criminal tendency, and some theories blamed deviant behavior on an abnormal shape in the frame or the skull. Frequently, the unusual were segregated from the rest of society in asylums. The rest of society maintained a fascination with the freakish element, and asylums often had viewing galleries so ordinary people could observe the activities and antics of the inmates.

7 The passage is reminiscent of Swift's *A Modest Proposal for Preventing the Children of the Poor People in Ireland from Being a Burden to Their Parents or Country, and for Making Them Beneficial to the Public* (1729), which gradually works its way round to

> I have been assured by a very knowing American of my acquaintance in London, that a young healthy child well nursed is at a year old a most delicious, nourishing and wholesome food, whether stewed, roasted, baked, or boiled, and I make no doubt that it will equally serve in a fricassee, or a ragout.

Swift then examines the economics of the proposal in great detail, in a masterpiece of satirical polemicism. As already noted, irony is a dangerous weapon, and Abbott's Swiftian jest ran a definite risk of backfiring. Swift was also a clergyman; perhaps literary clergy like to live dangerously.

when the frame is just beginning to set, and when the Medical Board has reported that recovery is improbable, I would suggest that the Irregular offspring be painlessly and mercifully consumed.[7]

§ 8. — *Of the Ancient Practice of Painting.*

IF MY Readers have followed me with any attention up to this point, they will not be surprised to hear that life is somewhat dull in Flatland. I do not, of course, mean that there are not battles, conspiracies, tumults, factions, and all those other phenomena which are supposed to make History interesting; nor would I deny that the strange mixture of the problems of life and the problems of Mathematics, continually inducing conjecture and giving the opportunity of immediate verification, imparts to our existence a zest which you in Spaceland can hardly comprehend. I speak now from the aesthetic and artistic point of view when I say that life with us is dull; aesthetically and artistically, very dull indeed.

How can it be otherwise, when all one's prospect, all one's landscapes, historical pieces, portraits, flowers, still life, are nothing but a single line, with no varieties except degrees of brightness and obscurity?

It was not always thus. Colour,[1] if Tradition speaks the truth, once for the space of half a dozen centuries or more, threw a transient splendour over[2] the lives of our ancestors in the re-

1 Although Abbott does not say so, color can be viewed as an extra dimension (or dimensions). Imagine, for instance, that an object's color can change continuously from red through purple to blue. Then, in addition to locating the position of that object in space (which requires two dimensions in Flatland and three in Spaceland), it is also necessary to locate the color of that object — which can be considered as a position in a "color space" running in the red-blue "direction"; see *Flatterland* chapter 4, "A Hundred and One Dimensions." There is no question that in this setting, color is independent of spatial location and can thus be considered an extra dimension. The idea is not new. In his 1927 *The Philosophy of Space and Time,* Hans Reichenbach says,

> Let us assume that the three dimensions of space are visualized in the customary fashion, and let us substitute a color for the fourth dimension. Every physical object is liable to changes in color as well as in position. An object might, for example, be capable of going through all shades from red through violet to blue. A physical interaction between any two bodies is possible only if they are close to each other in space as well as in color. Bodies of different colors would penetrate each other without interference....If we lock a number of [red] flies into a red glass globe, they may yet escape: they may change their color to blue and are then able to penetrate the red globe.

Artists have long located the colors perceived by the human eye in a two-dimensional "color triangle" whose vertices are "primary" colors and whose interior points are mixtures of these. The traditional primary colors are red, yellow, and blue, but we'll shortly see that other choices are more appropriate. In modern times, such concepts are vital to the computer graphics industry. Computer screen colors combine additively — the effect of a mixture depends on the sum of the component colors because the pixels (tiny fluorescent dots) that make up the screen *emit* light. The RGB color triangle for additive colors has vertices at Red, Green, and Blue. Combining

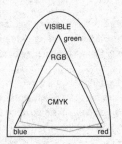

Figure 24 Color gamuts of RGB, CMYK, and visible light.

blue plus red (in equal amounts) gives magenta, red and green give yellow, and green and blue give cyan. All three together give white. Paints and inks, however, combine subtractively—the effect of a mixture depends on the differences between the component colors because paints *absorb* light (of all colors that differ from the one we see). The CMYK model employs Cyan, Magenta, and Yellow as its primary colors. A mixture of all three should in principle give black, but in practice the result is a muddy brown, so black (K) is added explicitly as a fourth component. Now a mixture of cyan and magenta gives blue, magenta and yellow give red, and yellow and cyan give green. The picture on your computer screen is created using the RGB system, but when you print it, your inkjet printer uses the CMYK system. This is one reason why digital photographs can look different on paper and on screen.

The issues here are complex, for some colors, such as brown, fall outside the standard color triangle. The *gamut* of a system is the range of colors it can produce. The CMYK gamut is smaller than the RGB gamut, which in turn is smaller than the full gamut of colors visible to the human eye (Figure 24). Other factors, such as brightness, introduce yet further dimensions into color vision. These extra dimensions are just as real as the traditional spatial dimensions. Indeed, the architecture of the visual system in the human brain is extremely complex because it has to take into account the additional dimensions of color and brightness. Reichenbach

motest ages. Some private individual—a Pentagon whose name is variously reported—having casually discovered the constituents of the simpler colours and a rudimentary method of painting, is said to have begun by decorating first his house, then his slaves, then his Father, his Sons, and Grandsons, lastly himself. The convenience as well as the beauty of the results commended themselves to all. Wherever Chromatistes,—for by that name the most trustworthy authorities concur in calling him,—turned his variegated frame, there he at once excited attention, and attracted respect. No one now needed to "feel" him; no one mistook his front for his back; all his movements were readily ascertained by his neighbours without the slightest strain on their powers of calculation; no one jostled him, or failed to make way for him; his voice was saved the labour of that exhausting utterance by which we colourless Squares and Pentagons are often forced to proclaim our individuality when we move amid a crowd of ignorant Isosceles.

The fashion spread like wildfire. Before a week was over, every Square and Triangle in the district had copied the example of Chromatistes, and only a few of the more conservative Pentagons still held out. A month or two found even the Dodecagons infected with the innovation. A year had not elapsed before the habit had spread to all but the very highest of the Nobility. Needless to say, the custom soon made its way from the district of Chromatistes to surrounding regions; and within two generations no one in all Flatland was colourless except the Women and the Priests.

Here Nature herself appeared to erect a barrier, and to plead against extending the innovation to these two classes. Many-sidedness was almost essential as a pretext for the Innovators. "Distinction of sides is intended by Nature to imply distinction of colours"—such was the sophism which in those days flew from mouth to mouth, converting whole towns at a time to the new culture. But manifestly to our Priests and Women this adage did not apply. The latter had only one side,[3] and therefore—plurally and pedantically speaking—*no sides*. The former—if at least they would assert their claim to be really and truly Circles,[4] and not mere high-class Polygons with an infinitely large number of infinitesimally small sides—were in the habit of boasting (what Women confessed and deplored) that they also had no sides, being blessed with a perimeter of one line, or, in other words, a Circumference. Hence it came to pass that these two Classes could see no force in the so-called axiom about "Distinction of Sides implying Distinction of Colour;" and when all others had succumbed to the fascinations of corporal decoration, the Priests and the Women alone still remained pure from the pollution of paint.

Immoral, licentious, anarchical, unscientific—call them by what names you will—yet, from an

chooses a one-dimensional subspace (red-violet-blue) of the two-dimensional color triangle to illustrate a fourth dimension. If he had chosen the entire spectrum of colors, he would have been visualizing a five-dimensional space.

2 Here the first edition has "... threw a transient charm upon ..."

3 This sentence is largely a play on words, and its mathematical meaning is delicate. Abbott discusses coloring polygons on the unstated assumption that each "side" (meaning "edge") gets a single color. There is, of course, no necessity to do this: The Flatland equivalent of stripes, or variegation, is straightforward. However, he is trying to set up a key narrative point: the dangerous equivalence of Women and Priests. To do this, he attempts to use a play on words to establish in the reader's mind that both Women and Priests have no sides. A side of a polygon is a straight line segment, and a Flatland Woman is a single straight line segment — hence "one side," hence (with a pointed remark about plurals) no sides. Similarly, a circle has a curved circumference, also "no sides." However, a line segment actually has two sides (left and right, so to speak), so — provided that the paint can be attached to the surface rather than pervading the whole line like a dye — Women could be painted with two colors. Indeed, a two-sided polygon (2-gon or digon) is a line segment described *twice,* which would be consistent with two distinct colors.

4 The wordplay here works only if the Circles — n-gons for very large n — can be considered true circles. Strictly speaking, this is not the case, but it would be a shame to let mere facts spoil a fantasy story, so Abbott slides neatly over the difficulty. In conventional mathematics, a circle is the *limit* of a sequence of n-gons as n tends to infinity, but it is not itself a polygon.

5 Here Abbott makes a rare slip, for such a sight could never be seen in the Flatland fog. Admittedly, the fog "prevails during the greater part of the year in all parts save the torrid zones" and so might not be present during this particular spectacle, but such spectacles must have been common because Abbott tells us that "How great and glorious the sensuous development of these days must have been is in part indicated by the very language and vocabulary of the period." Rare spectacles would not have had such a marked influence on day-to-day language.

aesthetic point of view, those ancient days of the Colour Revolt were the glorious childhood of Art in Flatland—a childhood, alas, that never ripened into manhood, nor even reached the blossom of youth. To live was then in itself a delight, because living implied seeing. Even at a small party, the company was a pleasure to behold; the richly varied hues of the assembly in a church or theatre are said to have more than once proved too distracting for our greatest teachers and actors; but most ravishing of all is said to have been the unspeakable magnificence of a military review.

The sight of a line of battle[5] of twenty thousand Isosceles suddenly facing about, and exchanging the sombre black of their bases for the orange and purple of the two sides including their acute angle; the militia of the Equilateral Triangles tricoloured in red, white, and blue; the mauve, ultra-marine, gamboge, and burnt umber of the Square artillerymen rapidly rotating near their vermilion guns; the dashing and flashing of the five-coloured and six-coloured Pentagons and Hexagons careering across the field in their offices of surgeons, geometricians and aides-de-camp—all these may well have been sufficient to render credible the famous story how an illustrious Circle, overcome by the artistic beauty of the forces under his command, threw aside his marshal's bâton and his royal crown, exclaiming that he henceforth exchanged them for the artist's pencil. How great and glorious the sensuous development of these days must have been is in part indicated by the very language and vocabulary of the period. The commonest utterances of the commonest citizens in the time of the Colour Revolt seem to have been suffused with a richer tinge of word or thought; and to that era we are even now indebted for our finest poetry and for whatever rhythm still remains in the more scientific utterance of these modern days.

1 The allegedly destructive influence of popular culture on the classics — dumbing down — was clearly perceived as a problem in Victorian times too.

§ 9. — Of the Universal Colour Bill.

But meanwhile the intellectual Arts were fast decaying.[1]

The Art of Sight Recognition, being no longer needed, was no longer practised; and the studies of Geometry, Statics, Kinetics, and other kindred subjects, came soon to be considered superfluous, and fell into disrespect and neglect even at our University. The inferior Art of Feeling speedily experienced the same fate at our Elementary Schools. Then the Isosceles classes, asserting that the Specimens were no longer used nor needed, and refusing to pay the customary tribute from the Criminal classes to the service of Education, waxed daily more numerous and more insolent on the strength of their immunity from the old burden which had formerly exercised the twofold wholesome effect of at once taming their brutal nature and thinning their excessive numbers.

Year by year the Soldiers and Artisans began more vehemently to assert — and with increasing truth — that there was no great difference between them and the very highest class of Polygons, now that they were raised to an equality with the latter, and enabled to grapple with all the difficulties and solve all the problems of life, whether Statical or Kinetical, by the simple process of Colour Recognition. Not content with the natural neglect into which Sight Recognition was falling, they began boldly to demand the legal prohibition of all "monopolizing and aristocratic Arts" and the consequent abolition of all endowments for the studies of Sight Recognition, Mathematics, and Feeling. Soon, they began to insist that inasmuch as Colour, which was a second Nature, had destroyed the need of aristocratic distinctions, the Law should follow in the same path, and that henceforth all individuals and all classes should be recognized as absolutely equal and entitled to equal rights.

Finding the higher Orders wavering and undecided, the leaders of the Revolution advanced still further in their requirements, and at last demanded that all classes alike, the Priests and the Women not excepted, should do homage to Colour by submitting to be painted. When it was objected that Priests and Women had no sides, they retorted that Nature and Expediency concurred in dictating that the front half of every human being (that is to say, the half containing his eye and mouth) should be distinguishable from his hinder half. They therefore brought before a general and extraordinary Assembly of all the States of Flatland a Bill proposing that in every Woman the half containing the eye and mouth should be coloured red, and the other half green. The Priests were to be painted in the same way, red being applied to that semicircle in which the eye and mouth formed the middle point; while the other or hinder semicircle was to be coloured green.

There was no little cunning in this proposal, which indeed emanated not from any Isosceles—for no being so degraded would have had angularity enough to appreciate, much less to devise, such a model of state-craft—but from an Irregular Circle who, instead of being destroyed in his childhood, was reserved by a foolish indulgence to bring desolation on his country and destruction on myriads of his followers.

On the one hand the proposition was calculated to bring the Women in all classes over to the side of the Chromatic Innovation. For by assigning to the Women the same two colours as were assigned to the Priests, the Revolutionists thereby ensured that, in certain positions, every Woman would appear like a Priest, and be treated with corresponding respect and deference—a prospect that could not fail to attract the Female Sex in a mass.

But by some of my Readers the possibility of the identical appearance of Priests and Women, under the new Legislation, may not be recognized; if so, a word or two will make it obvious.

Imagine a woman duly decorated, according to the new Code; with the front half (*i.e.* the half containing eye and mouth) red, and with the hinder half green. Look at her from one side. Obviously you will see a straight line, *half red, half green.*

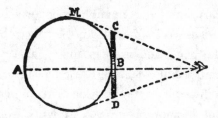

Now imagine a Priest, whose mouth is at M, and whose front semicircle (AMB) is consequently coloured red, while his hinder semicircle is green; so that the diameter AB divides the green from the red. If you contemplate the Great Man so as to have your eye in the same straight line as his dividing diameter (AB), what you will see will be a straight line (CBD), of which *one half* (CB) *will be red, and the other* (BD) *green*. The whole line (CD) will be rather shorter perhaps than that of a full-sized Woman, and will shade off more rapidly towards its extremities; but the identity of the colours would give you an immediate impression of identity of Class, making you neglectful of other details. Bear in mind the decay of Sight Recognition which threatened society at the time of the Colour Revolt; add too the certainty that Women would speedily learn[2] to shade off their extremities so as to imitate the Circles; it must then be surely obvious to you, my dear Reader, that the Colour Bill placed us under a great danger of confounding a Priest with a young Woman.

How attractive this prospect must have been to the Frail Sex may readily be imagined. They anticipated with delight the confusion that would

2 This entire passage imputes far more intelligence to Flatland women than the original discussion of their intellectual capacities would allow. It seems unlikely that this was an oversight. Abbott had an enlightened view of women's education: He was convinced that they could cope with topics far more intellectually demanding than sewing, cooking, and conversation. It was therefore natural for him to drop hints that Flatland's women had a lot more going for them than their menfolk habitually assumed. Elsewhere he tells of Flatland men who pretend to value women for their emotional attributes, but only when the women are present. In solely male company, they change their tune. Here we see that Flatland's women are also capable of misleading the opposite sex about their true nature. "Men are from Mars, women are from Venus" — so it was in Abbott's day, and so he wrote it in *Flatland*.

ensue. At home they might hear political and ecclesiastical secrets intended not for them but for their husbands and brothers, and might even issue commands in the name of a priestly Circle; out of doors the striking combination of red and green, without addition of any other colours, would be sure to lead the common people into endless mistakes, and the Women would gain whatever the Circles lost, in the deference of the passers by. As for the scandal that would befall the Circular Class if the frivolous and unseemly conduct of the Women were imputed to them, and as to the consequent subversion of the Constitution, the Female Sex could not be expected to give a thought to these considerations. Even in the households of the Circles, the Women were all in favour of the Universal Colour Bill.

The second object aimed at by the Bill was the gradual demoralization of the Circles themselves. In the general intellectual decay they still preserved their pristine clearness and strength of understanding. From their earliest childhood, familiarized in their Circular households with the total absence of Colour, the Nobles alone preserved the Sacred Art of Sight Recognition, with all the advantages that result from that admirable training of the intellect. Hence, up to the date of the introduction of the Universal Colour Bill, the Circles had not only held their own, but even increased their lead of the other classes by abstinence from the popular fashion.

Now therefore the artful Irregular whom I described above as the real author of this diabolical Bill, determined at one blow to lower the status of the Hierarchy by forcing them to submit to the pollution of Colour, and at the same time to destroy their domestic opportunities of training in the Art of Sight Recognition, so as to enfeeble their intellects by depriving them of their pure and colourless homes. Once subjected to the chromatic taint, every parental and every childish Circle would demoralize each other. Only in discerning between the Father and the Mother would the Circular infant find problems for the exercise of its understanding—problems too often likely to be corrupted by maternal impostures with the result of shaking the child's faith in all logical conclusions. Thus by degrees the intellectual lustre of the Priestly Order would wane, and the road would then lie open for a total destruction of all Aristocratic Legislature and for the subversion of our Privileged Classes.[3]

3 The City of London School, where Abbott worked for most of his professional life, was built on the site of Honey Lane Market in central London, not far from Saint Paul's Cathedral. A mile or so to the east lies the British Museum, where Karl Heinrich Marx (1818–1883) spent much of his time from 1849 onward. A year earlier, he and Friedrich Engels (1820–1895) had published *Manifest der Kommunistischen Partei* (*Manifesto of the Communist Party,* but usually translated as *The Communist Manifesto*). In 1867 (revised in 1885 and again in 1894) Marx published *Das Kapital* (*Capital,* in the sense of "capitalism"). Abbott could scarcely have been unaware of these developments, though his own politics, as a devotee of the Broad Church, was akin to what we now call social democracy. Naturally, the English privileged classes did not approve of Marxism or socialism.

1 Here Abbott is commenting on the proliferation of "universal rights" that began in the Victorian era. This was a time when the English social system was struggling toward greater equality — first for the common man and then for women, be they noble or common. Were it not for the rigidly conservative views of Queen Victoria, that goal might have been attained much sooner.

In 1832 the class of men able to vote in elections to Parliament was widened significantly, and the Reform Bill of 1867 widened it further, but not to women. The 1792 publication of *A Vindication of the Rights of Woman* by Mary Wollstonecraft (1759–1797) had given birth to what became known later as Woman Suffrage, and later still as the Women's Rights or Feminist Movement. In the 1870s Parliament received a petition with *3 million signatures* (the population of the United Kingdom was then about 28 million) urging that women be allowed to vote, but the prime ministers of the day — Gladstone and Disraeli — dared not oppose Her Majesty, so the issue was shelved. However, in 1869 some women were allowed to vote in municipal (local) elections. In 1897 all of the relevant organizations merged into a single National Union of Woman Suffrage Societies. Emmeline Pankhurst (1858–1928), her daughter Christabel (1880–1958), and the publicity stunts of the "suffragettes" kept the political pot boiling. After World War I, in which women had done large numbers of jobs previously restricted to men (such as manufacturing munitions), opposition evaporated. An act of Parliament extending the vote to women aged thirty or over passed in the House of Commons in 1917 and was ratified in the second chamber, the House of Lords, in 1918. In 1928 the voting age was reduced to twenty-one, the same as for men. (It is now eighteen.) Similar events unfolded in many other countries, notably the United States, beginning with Lucretia Mott (1793–1880) and Elizabeth Cady Stanton (1815–1902), whose advocacy of the antislavery movement led them into a general fight for women's rights. The first country

§ 10. — *Of the Suppression of the Chromatic Sedition.*

THE AGITATION for the Universal Colour Bill[1] continued for three years; and up to the last moment of that period it seemed as though Anarchy were destined to triumph.

A whole army of Polygons, who turned out to fight as private soldiers, was utterly annihilated by a superior force of Isosceles Triangles — the Squares and Pentagons meanwhile remaining neutral. Worse than all, some of the ablest Circles fell a prey to conjugal fury. Infuriated by political animosity, the wives in many a noble household wearied their lords with prayers to give up their opposition to the Colour Bill; and some, finding their entreaties fruitless, fell on and slaughtered their innocent children and husband, perishing themselves in the act of carnage. It is recorded that during that triennial agitation no less than twenty-three Circles[2] perished in domestic discord.

Great indeed was the peril. It seemed as though the Priests had no choice between submission and extermination; when suddenly the course of events was completely changed by one of those picturesque incidents which Statesmen ought never to neglect, often to anticipate, and sometimes perhaps to originate, because of the absurdly disproportionate power with which they appeal to the sympathies of the populace.

It happened that an Isosceles of a low type, with a brain little if at all above four degrees—accidentally dabbling in the colours of some Tradesman whose shop he had plundered—painted himself, or caused himself to be painted (for the story varies) with the twelve colours of a Dodecagon.[3] Going into the Market Place he accosted in a feigned voice a maiden, the orphan daughter of a noble Polygon, whose affection in former days he had sought in vain; and by a series of deceptions—aided, on the one side, by a string of lucky accidents too long to relate, and, on the other, by an almost inconceivable fatuity and neglect of ordinary precautions on the part of the relations of the bride—he succeeded in consummating the marriage. The unhappy girl committed suicide on discovering the fraud to which she had been subjected.

When the news of this catastrophe spread from State to State the minds of the Women were violently agitated. Sympathy with the miserable victim and anticipations of similar deceptions for themselves, their sisters, and their daughters, made them now regard the Colour Bill in an entirely new aspect. Not a few openly avowed themselves converted to antagonism; the rest needed

to give women the vote was New Zealand in 1893, followed by Australia in 1902.

Victorian intellectuals would have spent hours discussing such issues. Abbott was personally acquainted with many politicians and other leading figures of his time, and no doubt he saw *Flatland* as an opportunity to put some of his own views into print.

2 In nongeometric contexts, Abbott employs nonconstructible numbers — that is, numbers such that the polygon with that many sides cannot be constructed using ruler and compasses (and therefore makes no appearance in Euclid).

3 This is a twelve-sided polygon. The first edition of *Flatland* (and some modern editions, such as the Shambhala one, which appears to have been based on the first edition) has "dodecahedron." This is a *solid* with twelve pentagonal faces.

only a slight stimulus to make a similar avowal. Seizing this favourable opportunity, the Circles hastily convened an extraordinary Assembly of the States; and besides the usual guard of Convicts, they secured the attendance of a large number of reactionary Women.

Amidst an unprecedented concourse, the Chief Circle of those days—by name Pantocyclus—arose to find himself hissed and hooted by a hundred and twenty thousand Isosceles. But he secured silence by declaring that henceforth the Circles would enter on a policy of Concession; yielding to the wishes of the majority, they would accept the Colour Bill. The uproar being at once converted to applause, he invited Chromatistes, the leader of the Sedition, into the centre of the hall, to receive in the name of his followers the submission of the Hierarchy. Then followed a speech, a masterpiece of rhetoric, which occupied nearly a day in the delivery, and to which no summary can do justice.

With a grave appearance of impartiality he declared that as they were now finally committing themselves to Reform or Innovation, it was desirable that they should take one last view of the perimeter of the whole subject, its defects as well as its advantages. Gradually introducing the mention of the dangers to the Tradesmen, the Professional Classes and the Gentlemen, he silenced the rising murmurs of the Isosceles by reminding them that, in spite of all these defects, he was willing to accept the Bill if it was approved by the majority. But it was manifest that all, except the Isosceles, were moved by his words and were either neutral or averse to the Bill.

Turning now to the Workmen he asserted that their interests must not be neglected, and that, if they intended to accept the Colour Bill, they ought at least to do so with full view of the consequences. Many of them, he said, were on the point of being admitted to the class of the Regular Triangles; others anticipated for their children a distinction they could not hope for themselves. That honourable ambition would now have to be sacrificed. With the universal adoption of Colour, all distinctions would cease; Regularity would be confused with Irregularity; development would give place to retrogression; the Workman would in a few generations be degraded to the level of the Military, or even the Convict Class; political power would be in the hands of the greatest number, that is to say the Criminal Classes, who were already more numerous than the Workmen, and would soon out-number all the other Classes put together when the usual Compensative Laws of Nature were violated.

A subdued murmur of assent ran through the ranks of the Artisans, and Chromatistes, in alarm, attempted to step forward and address them. But he found himself encompassed with guards and forced to remain silent while the Chief Circle in a few impassioned words made a final appeal to the

4 The Victorians were no strangers to the use of military force to suppress the rights of subject peoples, and the best-known instance is the Indian Mutiny of 1857. India had in some sense been considered British territory since 1600, when the East India Company was given monopoly trading rights. In 1818 the British Empire *in* India became the British Empire *of* India; about half of India was ruled by Indians but was stripped of any serious military power and was split into about 360 politically independent units. In 1857 Indian troops (sepoys) under the East India Company embarked on a widespread rebellion against British rule. This is sometimes called the Sepoy Mutiny, but it went much deeper; in particular, the Hindu religion was being challenged by Christian missionaries, causing considerable resentment. The new Enfield rifle offered a pretext for revolt: Loading it necessitated biting off the ends of cartridges, and the sepoys believed — apparently with some justification — that the grease used as lubrication for the cartridges was made from a mixture of pig and cow fat. This was therefore an insult both to Muslims and to Hindus. In April 1857, sepoys at Meerut refused to use the cartridges and were jailed for lengthy terms, in irons. On May 10 their comrades mutinied, shot their British officers, and marched to Delhi. There they persuaded the former Mughal emperor Bahâdur Shâh II (1775–1862) to be named leader of the revolt, which spread to Agrâ, Cawnpore, and Lucknow. Fighting went on until peace was officially restored on 8 July 1858. There were atrocities on both sides, but British reprisals far outweighed the actions that caused them. Hundreds of sepoys were shot from cannons, and thousands of civilians were killed after perfunctory trials or no trials at all. Despite Bahâdur Shâh's age — he was over eighty — and the fact that he had been forced to accept nominal leadership, he and his family were exiled to Burma (now Myanmar). Britain blamed the East India Company for the mess, although British troops had been trampling all over the subcontinent for centuries,

Women, exclaiming that, if the Colour Bill passed, no marriage would henceforth be safe, no woman's honour secure; fraud, deception, hypocrisy would pervade every household; domestic bliss would share the fate of the Constitution and pass to speedy perdition. "Sooner than this," he cried, "Come death."

At these words, which were the preconcerted signal for action, the Isosceles Convicts fell on and transfixed the wretched Chromatistes; the Regular Classes, opening their ranks, made way for a band of Women who, under direction of the Circles, moved, back foremost, invisibly and unerringly upon the unconscious soldiers; the Artisans, imitating the example of their betters, also opened their ranks. Meantime bands of Convicts occupied every entrance with an impenetrable phalanx.

The battle, or rather carnage, was of short duration.[4] Under the skilful generalship of the Circles almost every Woman's charge was fatal and very many extracted their sting uninjured, ready for a second slaughter. But no second blow was needed; the rabble of the Isosceles did the rest of the business for themselves. Surprised, leader-less, attacked in front by invisible foes, and finding egress cut off by the Convicts behind them, they at once — after their manner — lost all presence of mind, and raised the cry of "treachery." This sealed their fate. Every Isosceles now saw and felt a foe in every other. In half an hour not one of that vast multitude was living; and the fragments of seven score thousand of the Criminal Class slain by one another's angles attested the triumph of Order.

The Circles delayed not to push their victory to the uttermost. The Working Men they spared but decimated. The Militia of the Equilaterals was at once called out; and every Triangle suspected of Irregularity on reasonable grounds, was destroyed by Court Martial, without the formality of exact measurement by the Social Board. The homes of the Military and Artisan classes were inspected in a course of visitations extending through upwards of a year; and during that period every town, village, and hamlet was systematically purged of that excess of the lower orders which had been brought about by the neglect to pay the tribute of Criminals to the Schools and University, and by the violation of the other natural Laws of the Constitution of Flatland. Thus the balance of classes was again restored.

Needless to say that henceforth the use of Colour was abolished, and its possession prohibited. Even the utterance of any word denoting Colour, except by the Circles or by qualified scientific teachers, was punished by a severe penalty. Only at our University in some of the very highest and most esoteric classes—which I myself have never been privileged to attend—it is understood that the sparing use of Colour is still sanctioned for the purpose of illustrating some of the deeper

and in 1858 the Government of India Act transferred power over India to the government of Queen Victoria. In 1876 she took on the title Empress of India.

5 Today's mathematicians, faced with the need to visualize complicated multidimensional geometry, often resort to the same remedy as the Flatlanders. Color can be used to distinguish different features of a complicated diagram with many overlapping components, making it easy to focus on any particular one — in effect imagining a "section" of the diagram at the appropriate level in the color dimension by ignoring all other colors. Or, especially in applications to science and commerce, the values of certain variables — "dimensions" — can be color-coded. In representations of the celebrated Mandelbrot set, a fractal introduced by Benoit Mandelbrot and one of the most complex forms ever specified by a simple equation, colors encode how rapidly a moving point escapes from some given region. See *Flatterland* chapter 5, "One and a Quarter Dimensions," and *The Beauty of Fractals* by Heinz-Otto Peitgen and Peter H. Richter. In *Symmetry in Chaos*, by Michael J. Field and Martin Golubitsky, color is used to encode the probability that a moving point lies in a given region of a dynamical attractor. The latter two of these books contain a wealth of beautiful color images.

problems of mathematics.[5] But of this I can only speak from hearsay.

Elsewhere in Flatland, Colour is now non-existent. The art of making it is known to only one living person, the Chief Circle for the time being; and by him it is handed down on his death-bed to none but his Successor. One manufactory alone produces it; and, lest the secret should be betrayed, the Workmen are annually consumed, and fresh ones introduced. So great is the terror with which even now our Aristocracy looks back to the far-distant days of the agitation for the Universal Colour Bill.

§ 11. — *Concerning our Priests.*

1 Abbott is not really interested in how a two-dimensional universe would really *work*; he needs it merely as a vehicle for popularizing the fourth dimension (and taking a sideswipe at Victorian society along the way). Recall that the challenge of making sense of two-dimensional physics was taken up by Hinton and Dewdney.

IT IS high time that I should pass from these brief and discursive notes about things in Flatland to the central event of this book, my initiation into the mysteries of Space. *That* is my subject; all that has gone before is merely preface.

For this reason I must omit many matters[1] of which the explanation would not, I flatter myself, be without interest for my Readers: as for example, our method of propelling and stopping ourselves, although destitute of feet; the means by which we give fixity to structures of wood, stone, or brick, although of course we have no hands, nor can we lay foundations as you can, nor avail ourselves of the lateral pressure of the earth; the manner in which the rain originates in the intervals between our various zones, so that the northern regions do not intercept the moisture from falling on the southern; the nature of our hills and mines, our trees and vegetables, our seasons and harvests; our Alphabet and method of writing, adapted to our linear tablets; these and a hundred other details of our physical existence I must pass over, nor do I mention them now except to indicate to my readers that their omission proceeds

2 Abbott needs to remind readers of the true nature of mathematical circles because he has already told them that Flatlanders move up the social scale, from generation to generation, when their descendants gain extra sides. This process can never produce a true mathematical circle, only an approximation to one. In conventional mathematics, the term polygon always implies a finite number of straight sides, so a circle cannot be a polygon. However, in *nonstandard analysis,* a recent development based on new ideas in mathematical logic, a circle can rigorously be considered to be a polygon with an infinite number of infinitesimal sides.

not from forgetfulness on the part of the author, but from his regard for the time of the Reader.

Yet before I proceed to my legitimate subject some few final remarks will no doubt be expected by my Readers upon those pillars and mainstays of the Constitution of Flatland, the controllers of our conduct and shapers of our destiny, the objects of universal homage and almost of adoration: need I say that I mean our Circles or Priests?

When I call them Priests, let me not be understood as meaning no more than the term denotes with you. With us, our Priests are Administrators of all Business, Art, and Science; Directors of Trade, Commerce, Generalship, Architecture, Engineering, Education, Statesmanship, Legislature, Morality, Theology; doing nothing themselves, they are the Causes of everything worth doing, that is done by others.

Although popularly everyone called a Circle is deemed a Circle, yet among the better educated Classes it is known that no Circle is really a Circle,[2] but only a Polygon with a very large number of very small sides. As the number of the sides increases, a Polygon approximates to a Circle; and, when the number is very great indeed, say for example three or four hundred, it is extremely difficult for the most delicate touch to feel any polygonal angles. Let me say rather, it *would* be difficult: for, as I have shown above, Recognition by Feeling is unknown among the highest society, and to *feel* a Circle would be considered a most audacious insult. This habit of abstention from Feeling in the best society enables a Circle the more easily to sustain the veil of mystery in which, from his earliest years, he is wont to enwrap the exact nature of his Perimeter or Circumference. Three feet being the average Perimeter[3] it follows that, in a Polygon of three hundred sides each side will be no more than the hundredth part of a foot in length, or little more than the tenth part of an inch; and in a Polygon of six or seven hundred sides the sides are little larger than the diameter of a Spaceland pin-head. It is always assumed, by courtesy, that the Chief Circle for the time being has ten thousand sides.

The ascent of the posterity of the Circles in the social scale is not restricted, as it is among the lower Regular classes, by the Law of Nature which limits the increase of sides to one in each generation. If it were so, the number of sides in a Circle would be a mere question of pedigree and arithmetic, and the four hundred and ninety-seventh descendant[4] of an Equilateral Triangle would necessarily be a Polygon with five hundred sides. But this is not the case. Nature's Law prescribes two antagonistic decrees affecting Circular propagation; first, that as the race climbs higher in the scale of development, so development shall proceed at an accelerated pace; second, that in the same proportion, the race shall become less fertile. Consequently in the home of a Polygon of four or five hundred sides it is rare to find a son; more than one is never seen. On the other hand the son of a five-hundred-sided Poly-

3 The perimeter of a circle of diameter 11 inches (the typical size in Flatland) is 11π inches, which amounts to approximately 34.5 inches — near enough to 3 feet. With 300 sides and a total length of 3 feet, each side will have length $^{36}/_{300} = 0.12$ inch — slightly more than a tenth of an inch, as Abbott states. With 600 sides, this figure reduces to 0.06 inch, which is near enough to the size of a pinhead.

4 In this passage, unlike the earlier one that refers to polygons with 3, 4, 5, 10, 12, 20, and 24 sides, the polygons are not constructible by ruler and compasses. The numbers of sides mentioned are (starting a few lines earlier and continuing to the end of the paragraph) 300, 600, 700, 10,000, 497, 500, 400, and 550. None of these satisfies the criterion for constructibility.

5 A Flatland analogue of gene therapy. Note that this treatment, when applied to a polygon with 200 or 300 sides, results in the skipping of 200 or 300 generations. Because the number of sides normally increases by 1 per generation, this statement is consistent with "doubles at a stroke."

gon has been known to possess five hundred and fifty, or even six hundred sides.

Art also steps in to help the process of the higher Evolution. Our physicians have discovered that the small and tender sides of an infant Polygon of the higher class can be fractured, and his whole frame re-set,[5] with such exactness that a Polygon of two or three hundred sides sometimes —by no means always, for the process is attended with serious risk—but sometimes overleaps two or three hundred generations, and as it were doubles at a stroke, the number of his progenitors and the nobility of his descent.

Many a promising child is sacrificed in this way. Scarcely one out of ten survives. Yet so strong is the parental ambition among those Polygons who are, as it were, on the fringe of the Circular class, that it is very rare to find a Nobleman of that position in society, who has neglected to place his firstborn in the Circular Neo-Therapeutic Gymnasium before he has attained the age of a month.

One year determines success or failure. At the end of that time the child has, in all probability, added one more to the tombstones that crowd the Neo-Therapeutic Cemetery; but on rare occasions a glad procession bears back the little one to his exultant parents, no longer a Polygon, but a Circle, at least by courtesy: and a single instance of so blessed a result induces multitudes of Polygonal parents to submit to similar domestic sacrifices, which have a dissimilar issue.

§ 12. — Of the Doctrine of our Priests.

1 In Flatland, one's social pedigree outweighs all else — talent, intelligence, ability — just as it did for the Victorians.

2 The story of Pantocyclus contradicts Abbott's earlier claims that Regularity is the natural order in Flatland. This is either a lapse or a clever way to remind his readers that hereditary privilege and ability do not always go together, despite conventional assumptions to the contrary.

As TO THE doctrine of the Circles it may briefly be summed up in a single maxim, "Attend to your Configuration."[1] Whether political, ecclesiastical, or moral, all their teaching has for its object the improvement of individual and collective Configuration — with special reference of course to the Configuration of the Circles, to which all other objects are subordinated.

It is the merit of the Circles that they have effectually suppressed those ancient heresies which led men to waste energy and sympathy in the vain belief that conduct depends upon will, effort, training, encouragement, praise, or anything else but Configuration. It was Pantocyclus — the illustrious Circle mentioned above, as the queller of the Colour Revolt — who first convinced mankind that Configuration makes the man; that if, for example, you are born an Isosceles with two uneven sides, you will assuredly go wrong unless you have them made even[2] — for which purpose you must go to the Isosceles Hospital; similarly, if you are a Triangle, or Square, or even a Polygon, born with any Irregularity, you must be taken to one of the Regular Hospitals to have your disease

3 In this one sentence, Abbott sums up a crucial issue of free will. Nowadays it is quite common for people accused of crimes to attribute their actions to some form of determinism (ranging from "I get aggressive when I'm drunk" to possession of a "gene for violence"). Leaving aside genuine and extreme cases of mental or genetic problems, Abbott's answer here seems entirely appropriate. Jack Cohen and I came to uncannily similar conclusions in *Figments of Reality* (independently, I'm sure, though it is *possible* that my earlier reading of *Flatland* lodged in my subconscious). In chapter 9, "We Wanted to Have a Chapter on Free Will, But We Decided Not to, So Here It Is" we wrote,

> In today's western culture, the question of free will is being obscured by an unfortunate side-effect of the simpleminded mental image of DNA as blueprint — a kind of genetic determinism. This adds genetic reinforcement to a growing cultural tendency for people not to accept responsibility for their own actions, which in a sense is an attempt to deny that they have free will....
>
> In Cyprus three soldiers who raped a young woman and beat her to death with a spade pled in their defence that they were so drunk that they did not know what they were doing. The Cypriot legal system wasn't impressed, but the defence lawyer clearly thought it was worth a try. In the very near future — if it has not happened already — a man will get drunk, kill somebody, and be acquitted because while drunk he is not responsible for his actions.
>
> In only the slightly more distant future, a man will kill somebody over a trivial argument, and be acquitted because he possesses a "gene for aggression." Those who *lack* a gene for aggression, yet fight back when attacked, will have no such excuse. If they are unfortunate enough also to possess a gene for rational decision-making, they will receive punitive sentences.... The underlying idea — that gross human characters such as aggression are somehow caused by a single segment of genetic code — stems from a

cured; otherwise you will end your days in the State Prison or by the angle of the State Executioner.

All faults or defects, from the slightest misconduct to the most flagitious crime, Pantocyclus attributed to some deviation from perfect Regularity in the bodily figure, caused perhaps (if not congenital) by some collision in a crowd; by neglect to take exercise, or by taking too much of it; or even by a sudden change of temperature, resulting in a shrinkage or expansion in some too susceptible part of the frame. Therefore, concluded that illustrious Philosopher, neither good conduct nor bad conduct is a fit subject, in any sober estimation, for either praise or blame. For why should you praise, for example, the integrity of a Square who faithfully defends the interests of his client, when you ought in reality rather to admire the exact precision of his right angles? Or again, why blame a lying, thievish Isosceles when you ought rather to deplore the incurable inequality of his sides?

Theoretically, this doctrine is unquestionable; but it has practical drawbacks. In dealing with an Isosceles, if a rascal pleads that he cannot help stealing because of his unevenness, you reply that for that very reason, because he cannot help being a nuisance to his neighbours, you, the Magistrate, cannot help sentencing him to be consumed[3] — and there's an end of the matter. But in little domestic difficulties, where the penalty of consumption, or death, is out of the question, this theory of

Configuration sometimes comes in awkwardly; and I must confess that occasionally when one of my own Hexagonal Grandsons pleads as an excuse for his disobedience that a sudden change of the temperature has been too much for his Perimeter, and that I ought to lay the blame not on him but on his Configuration, which can only be strengthened by abundance of the choicest sweetmeats, I neither see my way logically to reject, nor practically to accept, his conclusions.

For my own part, I find it best to assume that a good sound scolding or castigation has some latent and strengthening influence on my Grandson's Configuration; though I own that I have no grounds for thinking so. At all events I am not alone in my way of extricating myself from this dilemma; for I find that many of the highest Circles, sitting as Judges in law courts, use praise and blame towards Regular and Irregular Figures; and in their homes I know by experience that, when scolding their children, they speak about "right" or "wrong" as vehemently and passionately as if they believed that these names represented real existences, and that a human Figure is really capable of choosing between them.

Constantly carrying out their policy of making Configuration the leading idea in every mind, the Circles reverse the nature of that Commandment which in Spaceland regulates the relations between parents and children. With you, children are taught to honour their parents; with us—next

grievous misunderstanding of human development. The genome is more a recipe than a blueprint, and the ingredients and the skill of the cook are at least as important. A true human being has free will—in the only sense that matters, the relation of an individual to their culture—and is in control of its own destiny. A man who knows he gets aggressive when drunk, and kills while drunk, can try to excuse the murder—but he has no excuse for the drunkenness that he himself claims caused him to kill, because when he chose to get drunk he was sober. People who cannot control their tempers when drunk should consider themselves as being under a greater social obligation not to drink than those who can. Far from being a defence, drunkenness should compound the crime.... Apparently it's OK to disown responsibility for wielding a knife when drunk, but not for wielding a car.... Even if you think that all human behaviour is ultimately genetically determined, that still provides no reason for excusing murderers. If genes do actually correspond to characters, then as well as there being a gene for aggression there must also be a gene for controlling one's aggression, a gene for avoiding getting into tense situations, a gene for considering the effects of one's actions on other people ... a gene for taking responsibility. And if you are unfortunate enough to lack all of these genes, and I put you behind bars, don't complain: it's just that I happen to have a gene for incarcerating killers, it's not my fault.

4 This raises the whole question of sex in Flatland by making it clear that the reproductive system among the polygons is definitely sexual (and not, say, a process of budding or amoeba-like division). Flatland sex is one of those "many matters" that Abbott "must omit" — in this case, with true Victorian regard for propriety.

5 To all outward appearances, the female nobility of Flatland are indistinguishable from other females except by virtue of their family connections. Or, as Kipling put it in his barrack-room poem *The Ladies* in the *Pall Mall Gazette* of 2 May 1895,

> *The Colonel's Lady an' Judy O'Grady*
> *Are sisters under their skins!*

Abbott has little choice but to have females visibly indistinguishable because he has set them up as mere lines and the society has forbidden the use of color. However, he also tells us that the pedigree in Flatland certifies a degree of Regularity in offspring. This means that there must be hidden, "internal" differences in Flatland's women — whether or not we consider regularity to be a virtue — and these differences are to some extent inherited. The pedigree is a visible certification of these hidden differences. Abbott is satirizing the Victorian reliance on family trees as proof of "good breeding." But the edge is slightly taken off the satire because the Flatland pedigree is a genuine promise of future performance: regular children. Regularity may or may not be truly important in the Flatland scheme of things, but it is more tangible than inherited social class.

to the Circles, who are the chief object of universal homage — a man is taught to honour his Grandson, if he has one; or, if not, his Son. By "honour," however, is by no means meant "indulgence," but a reverent regard for their highest interests: and the Circles teach that the duty of fathers is to subordinate their own interests to those of posterity, thereby advancing the welfare of the whole State as well as that of their own immediate descendants.

The weak point in the system of the Circles —if a humble Square may venture to speak of anything Circular as containing any element of weakness — appears to me to be found in their relations with Women.

As it is of the utmost importance for Society that Irregular births should be discouraged, it follows that no Woman who has any Irregularities in her ancestry[4] is a fit partner for one who desires that his posterity should rise by regular degrees in the social scale.

Now the Irregularity of a Male is a matter of measurement; but as all Women are straight, and therefore visibly Regular so to speak, one has to devise some other means of ascertaining what I may call their invisible Irregularity, that is to say their potential Irregularities as regards possible offspring. This is effected by carefully-kept pedigrees,[5] which are preserved and supervised by the State; and without a certified pedigree no Woman is allowed to marry.

Now it might have been supposed that a Circle—proud of his ancestry and regardful for a posterity which might possibly issue hereafter in a Chief Circle—would be more careful than any other to choose a wife who had no blot on her escutcheon. But it is not so. The care in choosing a Regular wife appears to diminish as one rises in the social scale. Nothing would induce an aspiring Isosceles, who had hopes of generating an Equilateral Son, to take a wife who reckoned a single Irregularity among her Ancestors; a Square or Pentagon, who is confident that his family is steadily on the rise, does not inquire above the five-hundredth generation; a Hexagon or Dodecagon is even more careless[6] of the wife's pedigree; but a Circle has been known deliberately to take a wife who has had an Irregular Great-Grandfather, and all because of some slight superiority of lustre, or because of the charms of a low voice—which, with us, even more than with you, is thought "an excellent thing in Woman."

Such ill-judged marriages are, as might be expected, barren, if they do not result in positive Irregularity or in diminution of sides; but none of these evils have hitherto proved sufficiently deterrent. The loss of a few sides in a highly-developed Polygon is not easily noticed, and is sometimes compensated by a successful operation in the Neo-Therapeutic Gymnasium, as I have described above; and the Circles are too much disposed to acquiesce in infecundity as a Law of the superior

6 The first edition erroneously has "Dodecahedron" for "Dodecagon."

7 Part I of *Flatland* ("This World") ends with an extended statement extolling the virtues of better education of females, one of Abbott's professional interests. One aim of his 1883 *Hints on Home Teaching,* for instance, was to make it possible for girls to be educated in their own homes. At that time, the early education of boys and that of girls were very similar, except that for the girls, subjects such as needlework and housewifery were given precedence over arithmetic. The later education of the two sexes was very different, however. The boys (mainly upperclass, because children of the lower classes had to help the family by working for a living) were given a broad and deep education. As a sample, the 1836 curriculum of the City of London School — though admittedly this school was progressive for its time — lists reading, English grammar and composition, Latin, Greek, French, writing, arithmetic, bookkeeping, elements of mathematics and natural philosophy, geography, natural history, ancient and modern history, choral singing, and chemistry. Additional, optional courses included Hebrew, German, Spanish, Italian, and drawing. High-fliers could take Latin and Greek poetry, higher mathematics, physics, logic, and ethics. Often the boys attended a boarding school. There *were* boarding-schools for girls, but they were mostly "finishing schools" for upperclass young ladies, which taught them music, deportment, dancing, and French. Classics courses and mathematics more advanced than arithmetic (such as geometry and algebra) were entirely absent from the girls' curriculum.

8 This entrenched Victorian attitude persisted well beyond that period. Indeed, it is still widely held today, even though female mathematicians have proved themselves to be at least as capable as males. In 1859 J. S. Howson argued that girls should not be made to take examinations because of their "more excitable and sensitive constitutions." In 1911 the headmistress of Manchester High School for Girls stated that

development. Yet, if this evil be not arrested, the gradual diminution of the Circular class may soon become more rapid, and the time may be not far distant when, the race being no longer able to produce a Chief Circle, the Constitution of Flatland must fall.

One other word of warning[7] suggests itself to me, though I cannot so easily mention a remedy; and this also refers to our relations with Women. About three hundred years ago, it was decreed by the Chief Circle that, since women are deficient in Reason but abundant in Emotion, they ought no longer to be treated as rational, nor receive any mental education. The consequence was that they were no longer taught to read, nor even to master Arithmetic[8] enough to enable them to count the angles of their husband or children; and hence they sensibly declined during each generation in intellectual power. And this system of female non-education or quietism still prevails.

My fear is that, with the best intentions, this policy has been carried so far as to react injuriously on the Male Sex.

For the consequence is that, as things now are, we Males have to lead a kind of bi-lingual, and I may almost say bi-mental, existence. With Women, we speak of "love," "duty," "right," "wrong," "pity," "hope," and other irrational and emotional conceptions, which have no existence, and the fiction of which has no object except to control feminine exuberances; but among ourselves, and in our books, we have an entirely different vocabulary[9] and I may almost say, idiom. "Love" then becomes "the anticipation of benefits"; "duty" becomes "necessity" or "fitness"; and other words are correspondingly transmuted. Moreover, among Women, we use language implying the utmost deference for their Sex; and they fully believe that the Chief Circle Himself is not more devoutly adored by us than they are: but behind their backs they are both regarded and spoken of—by all except the very young—as being little better than "mindless organisms."

Our Theology also in the Women's chambers is entirely different from our Theology elsewhere.

Now my humble fear is that this double training, in language as well as in thought, imposes somewhat too heavy a burden upon the young, especially when, at the age of three years old, they are taken from the maternal care and taught to unlearn the old language—except for the purpose of repeating it in the presence of their Mothers and Nurses—and to learn the vocabulary and idiom of science. Already methinks I discern a weakness in the grasp of mathematical truth at the present time as compared with the more robust intellect of our ancestors three hundred years ago. I say nothing of the possible danger if a Woman should ever surreptitiously learn to read and convey to her Sex the result of her perusal of a single popular volume; nor of the possibility that the indiscretion or disobedience of some in-

even a moderate degree of success in mathematical study [by girls] ... can only be attained at an excessive cost in time, energy, and teaching power....Occasionally, there is so much strain and effort ... that, in consideration of her future health, the girl is obliged to give up the idea of going to college.

Yet, as A. G. Howson notes in his *A History of Mathematics Education in England* (1982), "Three years later the same girls were being asked to work long hours in ammunition factories or hospitals!"

9 Banchoff says that "Abbott was one of the first to recognize the implications of a 'two cultures' society" in which rational thinking is the territory of the men, and unquantifiable concepts such as love and loyalty are the sole responsibility of the women. However, the men pretend to share an interest in these female concerns when — and only when — women are present. A. Square's attitudes are only moderately enlightened: His main criticism of this kind of doublethink and doublespeak is that it imposes a terrible strain on the men. Elsewhere, Abbott wrote about the dangers of letting either the rational or the emotional side of one's personality dominate, implying that both men and women should combine these two ways of thinking and reacting. Abbott would have approved of today's "caring man," who lets his emotions show, and of the emancipated and technically skilled modern woman.

10 This final sentence of Part I sums up Abbott's views on female education with a direct appeal to the "highest Authorities." This alludes to a lengthy struggle that had been going on in Victorian education, in which the Misses Buss and Beale were prominent. The struggle had started with a few of the cheaper institutions, which taught girls reading, writing, arithmetic, grammar, geography, history, and maybe some French and music. Unmarried girls who had been educated at the cheaper schools usually became companions to the rich or governesses to their children. In 1848 the Governesses' Benevolent Institution, which aimed to help protect governesses (the protection was often needed), opened its own training college: Queen's College London. Two of the early students at Queen's College were the Misses Buss and Beale, who initiated major changes in women's secondary education. Miss Buss owned a private school that in 1850 became the North London Collegiate School for Ladies. In 1871 *Ladies* was changed to *Girls,* and the school was taken into public ownership. Other similar schools emerged, and the curriculum began to be determined by the public examinations, which were based on the existing boys' curriculum — itself in serious need of reform, but that's another matter. By 1881 Miss Buss wrote to Maria Georgia Grey (1816–1906) of Bishopsgate Training College that "There is now no such thing as a 'woman's education question'." This was perhaps overly optimistic: Mathematics, in particular, was considered an unsuitable subject for young ladies for many years after, along with all of the sciences that used it. This attitude has still not been fully eradicated.

A prevalent anonymous verse suggests that the students of Miss Buss and Miss Beale did not always appreciate the reduction of emotional content in their education. According to J. Kamm, in *How Different From Us: A Biography of Miss Buss and Miss Beale* (1958), an early version goes

fant Male might reveal to a Mother the secrets of the logical dialect. On the simple ground of the enfeebling of the Male intellect, I rest this humble appeal to the highest Authorities to reconsider the regulations of Female education.[10]

Miss Buss and Miss Beale
Cupid's darts do not feel;
They leave that to us
Poor Beale and poor Buss.

**A later version (attributed to "Sweet Seventeen")
ends with the lines**

Oh, how different from us
Are Miss Beale and Miss Buss.

PART II

OTHER WORLDS

"O brave new worlds,
That have such people in them!"[1]

1 The quotation is from Shakespeare's *The Tempest* (act 5, scene 1, line 183), where Miranda, daughter of Prospero, the rightful but exiled Duke of Milan, says, "O brave new world, that has such people in't!" referring to the first humans she has seen (apart from her father): They have been shipwrecked on the island where she lives, which is peopled by fantastical creatures. Clearly Abbott's misquotation is deliberate: His "brave new worlds" are the spaces of different dimensions, especially the third and higher. Abbott is expressing enthusiasm, as did Miranda, who prefaces this remark with "O, wonder! How many godly creatures are there here! How beauteous making is!" However, the same quotation inspired the ironic title of Aldous Huxley's *Brave New World,* about a totalitarian future in which babies are grown in machines, sex is purely recreational, drugs are officially sanctioned, and everyone is brainwashed into "knowing their place" in a rigid class system of alphas, betas, gammas, deltas, and epsilons. (Interestingly enough, Prospero shares Huxley's irony; his next line following Miranda's is "'Tis new to thee.")

§ 13.—*How I had a Vision of Lineland.*[1]

It was the last day but one of the 1999th year of our era,[2] and the first day of the Long Vacation.[3] Having amused myself till a late hour with my favourite recreation of Geometry,[4] I had retired to rest with an unsolved problem in my mind. In the night I had a dream.

I saw before me a vast multitude of small Straight Lines (which I naturally assumed to be Women)[5] interspersed with other Beings still smaller and of the nature of lustrous points—all moving to and fro in one and the same Straight Line, and, as nearly as I could judge, with the same velocity.

A noise of confused, multitudinous chirping or twittering issued from them at intervals as long as

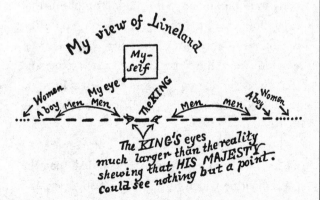

1 In order to strengthen his central analogy, Abbott embarks on a tour of alternative dimensions, starting with the one-dimensional world of Lineland. Here the King and Men are lines, and once again Women are one dimension lower — in this case zero-dimensional, mere points. By giving A. Square a *simpler* world to look down upon, Abbott sets the scene for A. Square's discovery that *his* own world is simpler to a three-dimensional observer. This paves the way for an attempt to convince hidebound Victorians that *their* three-dimensional world might be just a tiny part of a more complex, four-dimensional one. Lineland almost certainly derives from Hinton's 1880 pamphlet "What Is the Fourth Dimension?" In the 1907 *An Episode of Flatland,* Hinton returns to the idea with "the case of a being lower in the scale of space existence … confined altogether to a line."

2 In order to explain the context here, it is necessary to give away part of Abbott's plot. This requires the Sphere, his visitor from the third dimension, to appear before the high priesthood once every millennium. The priests *know* that the third dimension is real, but they want that information suppressed because it might damage their own power. Recall that Abbott belonged to the Broad Church, whose attitudes were democratic. The High Church, on the other hand, was elitist, fixated on ceremony, and determined to hold on to power — and so was the Roman Catholic Church. The Flatland priests are caricatures, either of the Anglican High Churchmen or of the cardinals and the pope of the Church of Rome.

Abbott's choice of date deserves discussion. In the Western world of Spaceland, there was (and still is) a widespread belief that the twentieth century ended on 31 December 1999 and that 1 January 2000 marked the start of the twenty-first. This belief was natural, given the numerology, and was encouraged by national governments with "Millennial" projects and events and by simple considerations of marketing (2000 is a more recognizable millennial brand than 2001). A peculiarity of the calendar, however, means that all centuries start in years that end 01 and all millennia begin in years that end 001. *The Guardian* newspaper acknowledged as much in its editorial for the edition of 1 January 2001, adding that

> a piece in *G2* [a subsection of the newspaper] pointed out last week [that the Victorians] saluted the end of the 19th Century and the birth of the 20th on the night of December 31 1900. The poet laureate, Alfred Austin, produced a (putrid) poem to mark the occasion.

The same source states that alone among all British municipalities, the town of Hythe (on the Kent coast opposite France) reserved its millennial celebrations for the end of 2000. This was thanks to Denise Rayner, who, armed only with yellowing copies of late-nineteenth-century newspapers, persuaded the town to adopt the resolution "that Hythe shows its independence and common sense by greeting the 21st Century and the third millennium on the correct date of January 1 2001 and not a year earlier." SF fans all knew that 2001 was the start of the third millennium because that is why Arthur C(harles) Clarke (1917–) called his book and movie (with producer Stanley Kubrick, 1928–1999, released in 1968) *2001: A Space Odyssey* and not *2000: A Space Odyssey*.

Why do we get ourselves into this kind of muddle? All calendric systems are compromises because the rotational period of the Earth ("day") is not some simple fraction of its period of revolution around the sun ("year"). Moreover, the period of revolution of the moon around the Earth ("month") is commensurate with neither. The calendar that is now used worldwide for civil purposes — though many nations retain different calendars for religious purposes — evolved from the Roman calendar. The Roman calendar was based on the phases of the moon, and it quickly got out of step with the seasons, which depend on the Earth's position relative to the sun. The Roman calendar in effect adopted a

they were moving; but sometimes they ceased from motion, and then all was silence.

Approaching one of the largest of what I thought to be Women, I accosted her, but received no answer. A second and a third appeal on my part were equally ineffectual. Losing patience at what appeared to me intolerable rudeness, I brought my mouth into a position full in front of her mouth so as to intercept her motion, and loudly repeated my question, "Woman, what signifies this concourse, and this strange and confused chirping, and this monotonous motion to and fro in one and the same Straight Line?"

"I am no Woman," replied the small Line: "I am the Monarch of the world. But thou, whence intrudest thou into my realm of Lineland?" Receiving this abrupt reply, I begged pardon if I had in any way startled or molested his Royal Highness; and describing myself as a stranger I besought the King to give me some account of his dominions. But I had the greatest possible difficulty in obtaining any information on points that really interested me; for the Monarch could not refrain from constantly assuming that whatever was familiar to him must also be known to me and that I was simulating ignorance in jest. However, by persevering questions I elicited the following facts:

It seemed that this poor ignorant Monarch— as he called himself—was persuaded that the Straight Line which he called his Kingdom, and in which he passed his existence, constituted the whole of the world, and indeed the whole of Space. Not being able either to move or to see, save in his Straight Line, he had no conception of anything out of it. Though he had heard my voice when I first addressed him, the sounds had come to him in a manner so contrary to his experience that he had made no answer, "seeing no man," as he expressed it, "and hearing a voice as it were from my own intestines." Until the moment when I placed my mouth in his World, he had neither seen me, nor heard anything except confused sounds beating against—what I called his side, but what he called his *inside* or *stomach*; nor had he even now the least conception of the region from which I had come. Outside his World, or Line, all was a blank to him; nay, not even a blank, for a blank implies Space; say, rather, all was non-existent.

His subjects—of whom the small Lines were men and the Points Women—were all alike confined in motion and eye-sight to that single Straight Line, which was their World. It need scarcely be added that the whole of their horizon was limited to a Point; nor could any one ever see anything but a Point. Man, woman, child, thing—each was a Point to the eye of a Linelander. Only by the sound of the voice could sex or age be distinguished. Moreover, as each individual occupied the whole of the narrow path, so to speak, which constituted his Universe, and no one could move to the right or left to make way for passers by, it

year that on average was 366 1/4 days long—one day too many — and by 50 B.C., spring had slipped 8 weeks and was starting in May. Julius Caesar (100? B.C.–44 B.C.) asked the Alexandrian astronomer Sosigenes (flourished 50 B.C.) for advice, and he suggested switching to the Egyptian year of 365 1/4 days. To return the seasons to their rightful places, the year 46 B.C. was accorded 445 days. To deal with that awkward 1/4 day, Sosigenes proposed to have a normal year that lasted 365 days but to add an extra "leap" day every fourth year (that extra day is now February 29, but originally it fell between February 23 and 24). In 44 B.C. further minor changes to the lengths of the months were made by Augustus Caesar (63 B.C.–A.D. 14), but this "Julian calendar" served humanity well for 1,500 years. However, the tropical year is actually 365.242199 days, not 365.25 — an error of 11 minutes 14 seconds per year. By 1545 the vernal equinox (the beginning of spring, used to work out the date of Easter) was wrong by 10 days, and the Council of Trent authorized Pope Paul III (1468–1549) to put matters right. Various remedies were proposed, but no action was taken until Pope Gregory XIII (1502–1585) was elected in 1572. Then the astronomer and Euclidean commentator Christopher Clavius (1537–1612) drafted a papal bull based on ideas of the astronomer Luigi Lilio. The bull, which appeared in 1582, assumed a year of 365.2422 days and adjusted the calendar by omitting leap years in all years that are multiples of 100, *except* that multiples of 400 remain leap years (as, for example, was the case in A.D. 2000). Ten days were omitted to return the vernal equinox to its correct position. The resulting "Gregorian calendar" is the one we use today. By the time it was introduced into England, 11 days had to be omitted, giving rise to the famous cries of "give us back our eleven days!" This complaint made more sense than it seems because payment of taxes depended on the boundaries between months.

The Gregorian calendar was also extrapolated

backward in time, with the year A.D. 1 marking the "official" year of Christ's birth and earlier dates being reckoned "before Christ" (B.C.). This is where the confusion about centuries and millennia arose, because the Gregorian calendar lacks a year 0: The year A.D. 1 is preceded by 1 B.C. (The Flatland calendar does have a year 0 and so has logical dates for millennia.) As a consequence, the first century of the Gregorian calendar was deemed to have begun on 1 January of the year 1 and to have ended on 31 December 100, and so on. The twentieth century therefore ran from 1 January 1901 to 31 December 2000, so the first day of the twenty-first century, and with it the third millennium, was 1 January 2001. Widespread references to this contention as "controversial" missed the point: It was both legally and calendrically *correct*. The Victorians knew their calendar, and they celebrated the start of the twentieth century at the beginning of 1901, *not* of 1900. Thanks to dumbing down, their successors 99 years later got it hopelessly wrong.

3 At Oxford and Cambridge universities, the Long Vacation runs for the summer months — roughly July to September, inclusive. The other two vacations, Christmas and Easter, are much shorter — hence the name. Abbott was an undergraduate at Cambridge and then a fellow and later an honorary fellow of St. John's College, Cambridge. He was also Hulsean lecturer at the university in 1876 and select preacher at Oxford University in 1877. However, the Flatlanders' Long Vacation starts on their equivalent of December 30. (Abbott does not specify their calendric system beyond allusions to year numbers.) Presumably, Abbott borrowed the phrase without being concerned about precise timing.

4 This is the first explicit statement that A. Square is good at mathematics; there are other hints later, too. We know that Abbott himself was competent at mathematics, including geometry, because he said so in 1894 at an Old Boys' reunion dinner:

followed that no Linelander could ever pass another. Once neighbours, always neighbours. Neighbourhood with them was like marriage with us. Neighbours remained neighbours till death did them part.

Such a life, with all vision limited to a Point, and all motion to a Straight Line, seemed to me inexpressibly dreary; and I was surprised to note the vivacity and cheerfulness of the King. Wondering whether it was possible, amid circumstances so unfavourable to domestic relations, to enjoy the pleasures of conjugal union, I hesitated for some time to question his Royal Highness on so delicate a subject; but at last I plunged into it by abruptly inquiring as to the health of his family. "My wives and children," he replied, "are well and happy."

Staggered at this answer — for in the immediate proximity of the Monarch (as I had noted in my dream before I entered Lineland) there were none but Men — I ventured to reply, "Pardon me, but I cannot imagine how your Royal Highness can at any time either see or approach their Majesties, when there are at least half a dozen intervening individuals, whom you can neither see through, nor pass by? Is it possible that in Lineland proximity is not necessary for marriage and for the generation of children?"

"How can you ask so absurd a question?" replied the Monarch. "If it were indeed as you suggest, the Universe would soon be depopulated.

No, no; neighbourhood is needless for the union of hearts; and the birth of children is too important a matter to have been allowed to depend upon such an accident as proximity. You cannot be ignorant of this. Yet since you are pleased to affect ignorance, I will instruct you as if you were the veriest baby in Lineland. Know, then, that marriages are consummated by means of the faculty of sound and the sense of hearing.

"You are of course aware that every Man has two mouths or voices—as well as two eyes—a bass at one and a tenor at the other of his extremities. I should not mention this, but that I have been unable to distinguish your tenor in the course of our conversation." I replied that I had but one voice, and that I had not been aware that his Royal Highness had two. "That confirms my impression," said the King, "that you are not a Man, but a feminine Monstrosity with a bass voice, and an utterly uneducated ear. But to continue.

"Nature having herself ordained that every Man should wed two wives——"[6] "Why two?" asked I. "You carry your affected simplicity too far," he cried. "How can there be a completely harmonious union without the combination of the Four in One, viz. the Bass and Tenor of the Man and the Soprano and Contralto of the two Women?" "But supposing," said I, "that a man should prefer one wife or three?" "It is impossible," he said; "it is as inconceivable as that two and one should make five, or that the human eye should see a

I can well remember being taken in the year 1850 by my father, with a number of young aspirants, to the City of London School … and being put through my paces by Dr. Mortimer [Rev. G. F. W. Mortimer, Headmaster from 1840 to 1865] … and I happened to reply, in answer to questions in mathematics … that I had learned the first book of Euclid.

Moreover, his Cambridge degree in classics required him to pass courses from the mathematics degree. As a pupil at the school he studied mathematics under Edkins, a brilliant though eccentric teacher. H. S. Fagan, who had been a boarder at the school, recalled him as a "… heroic man … who not content with turning out more single-figure wranglers than any other master of his day, was determined also to initiate the common herd of us into his mysteries." He also described one of Edkins's favorite punishments:

I remember [my friend and I] were ostracized [excluded from the class] all the while the class was doing geometrical conics. I led the rebellion, asserting that algebra was the proper language for conic sections…. [This from one of the "common herd"!] We would not learn, and when impositions were found unavailing, we were banished from the class…. [Impositions are a form of punishment involving copying out lengthy sections of an assigned text.] By and by, they began the calculus, and I who had read little of it by myself and found the enforced leisure of the mathematical hours [ostracism!] grow wearying, humbly petitioned to be allowed to join…. With rare magnanimity the master whom I had so long worried took me back unconditionally.

Edkins was easily angered; he had no feeling for discipline and was the constant butt of practical jokes. The boys would place rows of "skipjacks" in a classroom while they sat some distance away in perfect silence. These ingenious articles of torment were made from walnut shells and twisted rubber with a match inside, sealed with wax. When the wax

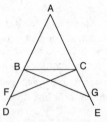

Figure 25 The *pons asinorum*: how to prove that the angles at the base of an isosceles triangle are equal.

melted, the device would leap into the air with a loud *plop!* Former pupil E. W. Emery relates an anecdote involving the diagram for Euclid's *Elements* [1,5], traditionally known as the *pons asinorum,* or "bridge of asses" (Figure 25). This is the theorem that the base angles of an isosceles triangle are equal, and it involves some auxiliary constructions and a series of arguments about congruent triangles. The figure looks a little like a bridge, and the Proposition was notorious as a major stumbling block for those with little mathematical talent. Every day, Edkins would draw this figure on the board "for the benefit of an unfortunate who could not learn it, and the daily flogging which followed till the ass had crossed the bridge." Edkins was absentminded and nearsighted: "On one occasion an umbrella was opened in the Sixth Class room. Edkins rushed into the room and hurled it out of the window.... Unfortunately it was his own umbrella, which he had left in the form-room and forgotten."

5 Actually, there is no very good reason why A. Square should make that assumption. Both polygons and line segments appear to his visual sense as lines. A line segment might be a Woman, but it might also be a Triangle, or a Square, seen from a direction in which only one face is visible. The assumption of Women sets the reader up for Abbott's surprise: The lines are actually Men, in a lower-dimensional world. However, he has not thought the Flatland point of view through fully here. (There is a

Straight Line." I would have interrupted him; but he proceeded as follows:

"Once in the middle of each week a Law of Nature compels us to move to and fro with a rhythmic motion of more than usual violence, which continues for the time you would take to count a hundred and one. In the midst of this choral dance, at the fifty-first pulsation, the inhabitants of the Universe pause in full career, and each individual sends forth his richest, fullest, sweetest strain. It is in this decisive moment that all our marriages are made. So exquisite is the adaptation of Bass to Treble, of Tenor to Contralto, that oftentimes the Loved Ones, though twenty thousand leagues away, recognize at once the responsive note of their destined Lover; and, penetrating the paltry obstacles of distance, Love unites the three. The marriage in that instant consummated results in a threefold Male and Female offspring which takes its place in Lineland."

"What! Always threefold?" said I. "Must one wife then always have twins?"

"Bass-voiced Monstrosity! yes," replied the King. "How else could the balance of the Sexes be maintained, if two girls were not born for every boy? Would you ignore the very Alphabet of Nature?" He ceased, speechless for fury; and some time elapsed before I could induce him to resume his narrative.

"You will not, of course, suppose that every bachelor among us finds his mates at the first wooing in this universal Marriage Chorus. On the contrary, the process is by most of us many times repeated. Few are the hearts whose happy lot it is at once to recognize in each other's voices the partner intended for them by Providence, and to fly into a reciprocal and perfectly harmonious embrace. With most of us the courtship is of long duration. The Wooer's voices may perhaps accord with one of the future wives, but not with both; or not, at first, with either; or the Soprano and Contralto may not quite harmonize. In such cases Nature has provided that every weekly Chorus shall bring the three Lovers into closer harmony. Each trial of voice, each fresh discovery of discord, almost imperceptibly induces the less perfect to modify his or her vocal utterance so as to approximate to the more perfect. And after many trials and many approximations, the result is at last achieved. There comes a day at last, when, while the wonted Marriage Chorus goes forth from universal Lineland, the three far-off Lovers suddenly find themselves in exact harmony, and, before they are aware, the wedded Triplet is rapt vocally into a duplicate embrace; and Nature rejoices over one more marriage and over three more births."

defense: Abbott's protagonist would be unlikely to see a whole series of Squares, *all* from a direction in which only one face is visible.)

6 Abbott's solution to the problem of maintaining a stable sex ratio in Lineland is elegant and unorthodox. In Spaceland, the same question arises in many species: Why is the sex ratio stably maintained at (roughly) 50:50? Given that the same male could impregnate many females, why does the ratio not become biased toward a predominance of females, on grounds of evolutionary efficiency? The "operational" answer is that (in humans, a typical case) sex is determined by two sex chromosomes (a chromosome is a package of genetic material). One sex chromosome is inherited from the father, the other from the mother. The female sex chromosome is called X, the male one Y. Normal females have sex chromosomes XX; normal males have XY. (There are abnormalities, such as XXY; we ignore these here.) Random choices of a single sex chromosome from each parent preserve the two possibilities XX and XY, and they also preserve the 50:50 ratio, assuming equal mortality rates, which is not quite correct. (The actual proportions in humans are 53 boys to 50 girls at birth, and the disparity that would be required to even up the numbers by puberty, thanks to the higher mortality of boys, is *more* than this, but the reasons are not understood.) The mother necessarily contributes an X. The father is equally likely to contribute an X, leading to XX in the offspring, or a Y, leading to XY. A more symmetric system, XX for female and YY for male, would also lead to equal proportions of XX and YY in the next generation, but there would also be (twice as many) cases of XY. The XY's could be eliminated by some mechanism, but the system that is actually used avoids such waste.

That is *how* the 50:50 proportion is maintained, but the question of why remains. The generally accepted answer was given by the geneticist R(onald) A(ylmer) Fisher (1890–1962). His argument is subtle and involves three successive generations. The main idea is that there is a stabilizing feedback effect. Assume that there exists some genetic mechanism to modify the sex ratio (if not, there is nothing to explain). If there are more than 50% females, it pays females to produce more sons; if there are fewer than 50% females, it pays females to produce more daughters. Pays here does not imply conscious knowledge: In the long run, evolution will select such strategies by trial and error.

To examine this in more detail: suppose that the female can choose the sex of each child at will. (Again this is not to be taken literally. It is just a convenient metaphor for a corresponding evolutionary possibility.) If we fix the total number of children, no change of strategy can affect how many children she has, but it *can* affect how many grandchildren she has. Specifically, if girls are rarer than boys, then she improves her chances of passing on her genes (embodied in her grandchildren) by having a girl, where the competition for mates is less. Similarly, if boys are rarer than girls, then having a boy improves her chances of passing on her genes.

There is, however, a problem with Fisher's explanation. It assumes that all males are equally likely to mate and that all (or at least a statistically representative sample of) offspring survive to reproduce as adults. Neither assumption is valid, yet the sex ratio remains close to 50:50 in innumerable species that violate Fisher's assumptions. In elephant seals, for example, perhaps 1 male in 20 actually sires off-spring — the dominant bull who maintains a harem of females. The average harem contains 20 females, however, so the 50:50 sex ratio is closely adhered to. Fisher's explanation does not apply directly, and it is not clear that any simple modification does either. Hence the puzzle remains, even though most geneticists are taught that it has long since been solved.

§ 14. — *How I vainly tried to explain the nature of Flatland.*

1 Again Abbott is setting his reader up for the dimensional analogy. A. Square observes Lineland from the outside, so what might observe Flatland from the outside? Spaceland, of course. But, more subtly, what might observe *Spaceland* from the outside? Why should the dimensions of space stop at three, when the analogy continues to work perfectly well?

THINKING THAT it was time to bring down the Monarch from his raptures to the level of common sense, I determined to endeavour to open up to him some glimpses of the truth, that is to say of the nature of things in Flatland. So I began thus: "How does your Royal Highness distinguish the shapes and positions of his subjects? I for my part noticed by the sense of sight, before I entered your Kingdom,[1] that some of your people are Lines and others Points, and that some of the Lines are larger——" "You speak of an impossibility," interrupted the King; "you must have seen a vision; for to detect the difference between a Line and a Point by the sense of sight is, as every one knows, in the nature of things, impossible; but it can be detected by the sense of hearing, and by the same means my shape can be exactly ascertained. Behold me—I am a Line, the longest in Lineland, over six inches of Space——" "Of Length," I ventured to suggest. "Fool," said he, "Space is Length. Interrupt me again, and I have done."

I apologized; but he continued scornfully, "Since

2 A clever use of the speed of sound to allow the King of Lineland to perceive his surroundings — and a gentle reminder to his Victorian readers that sound *has* a speed (about 1100 feet, or 340 meters, per second, depending on temperature and other factors). However, Abbott missed a chance to remind them of the Doppler effect, discovered in 1842 by Christian Doppler (1803–1853), whereby the perceived frequency of sound depends on the relative motion of the emitter and the receiver. This is why the sound of a passing ambulance suddenly becomes lower in pitch as it passes a stationary observer. When the ambulance is approaching the observer, more sound waves reach the observer per second than would be the case if the ambulance were also stationary — because later waves have a shorter distance to travel. The frequency of the sound thus appears higher than it would be if the sound were emitted by a stationary ambulance. Similarly, when the ambulance is moving away, fewer sound waves reach the observer per second than would be the case if the ambulance were also stationary — because later waves have a longer distance to travel. The frequency of the sound thus appears lower than it would be if the sound were omitted from a stationary ambulance.

Today the most important aspect of the Doppler effect is its relativistic analogue, which led to the discovery that the universe is expanding. Here, the effect occurs for light rather than sound. Light also has a finite speed, though it is *much* faster than sound: 300,000 kilometers (186,000 miles) per *second*. The Doppler effect for light causes a change of frequency, which is perceived as a change of *color*. Between 1912 and 1925, the American astronomer Vesto Slipher (1875–1969) discovered a "red shift" in the spectra of distant galaxies, which — thanks to the Doppler effect — implies that they are moving rapidly away from us. Another American astronomer, Edwin Powell Hubble (1889–1953), found a relationship between the size of the shift and the distance of the galaxy: More distant galaxies are reced-

you are impervious to argument, you shall hear with your ears how by means of my two voices I reveal my shape to my Wives, who are at this moment six thousand miles seventy yards two feet eight inches away, the one to the North, the other to the South. Listen, I call to them."

He chirruped, and then complacently continued: "My wives at this moment receiving the sound of one of my voices, closely followed by the other, and perceiving that the latter reaches them after an interval[2] in which sound can traverse 6.457 inches, infer that one of my mouths is 6.457 inches further from them than the other, and accordingly know my shape to be 6.457 inches. But you will of course understand that my wives do not make this calculation every time they hear my two voices. They made it, once for all, before we were married. But they *could* make it at any time. And in the same way I can estimate the shape of any of my Male subjects by the sense of sound."

"But how," said I, "if a Man feigns a Woman's voice with one of his two voices, or so disguises his Southern voice that it cannot be recognized as the echo of the Northern? May not such deceptions cause great inconvenience? And have you no means of checking frauds of this kind by commanding your neighbouring subjects to feel one another?" This of course was a very stupid question, for feeling could not have answered the purpose; but I asked with the view of irritating the Monarch, and I succeeded perfectly.

"What!" cried he in horror, "explain your meaning." "Feel, touch, come into contact," I replied. "If you mean by *feeling*," said the King, "approaching so close as to leave no space between two individuals, know, Stranger, that this offence is punishable in my dominions by death. And the reason is obvious. The frail form of a Woman, being liable to be shattered by such an approximation, must be preserved by the State; but since Women cannot be distinguished by the sense of sight from Men, the Law ordains universally that neither Man nor Woman shall be approached so closely as to destroy the interval between the approximator and the approximated.

"And indeed what possible purpose would be served by this illegal and unnatural excess of approximation which you call *touching*, when all the ends of so brutal and coarse a process are attained at once more easily and more exactly by the sense of hearing? As to your suggested danger of deception, it is non-existent: for the Voice, being the essence of one's Being, cannot be thus changed at will. But come, suppose that I had the power of passing through solid things, so that I could penetrate my subjects, one after another, even to the number of a billion, verifying the size and distance of each by the sense of *feeling*: how much time and energy would be wasted in this clumsy and inaccurate method! Whereas now, in one moment of audition, I take as it were the census and statistics, local, corporeal, mental and spiri-

ing faster. This result is now interpreted as evidence for the expansion of the universe. The mathematical details are different in the Newtonian and relativistic cases. In Newtonian physics, the perceived frequency is never less than half the true frequency, whereas in relativistic physics, the ratio can be as close as we like to zero, provided that the source travels at close enough to the speed of light. The difference between Newtonian and relativistic ratios becomes significant only when the source is traveling faster than about one-third the speed of light.

tual, of every living being in Lineland. Hark, only hark!"

So saying he paused and listened, as if in an ecstasy, to a sound which seemed to me no better than a tiny chirping from an innumerable multitude of lilliputian grasshoppers.

"Truly," replied I, "your sense of hearing serves you in good stead, and fills up many of your deficiencies. But permit me to point out that your life in Lineland must be deplorably dull. To see nothing but a Point! Not even to be able to contemplate a Straight Line! Nay, not even to know what a Straight Line is! To see, yet to be cut off from those Linear prospects which are vouchsafed to us in Flatland! Better surely to have no sense of sight at all than to see so little! I grant you I have not your discriminative faculty of hearing; for the concert of all Lineland which gives you such intense pleasure, is to me no better than a multitudinous twittering or chirping. But at least I can discern, by sight, a Line from a Point. And let me prove it. Just before I came into your kingdom, I saw you dancing from left to right, and then from right to left, with Seven Men and a Woman in your immediate proximity on the left, and eight Men and two Women on your right. Is not this correct?"

"It is correct," said the King, "so far as the numbers and sexes are concerned, though I know not what you mean by 'right' and 'left.' But I deny that you saw these things. For how could you see the Line, that is to say the inside, of any Man? But you must have heard these things, and then dreamed that you saw them. And let me ask what you mean by those words 'left' and 'right.' I suppose it is your way of saying Northward and Southward."

"Not so," replied I; "besides your motion of Northward and Southward, there is another motion which I call from right to left."

King. Exhibit to me, if you please, this motion from left to right.

I. Nay, that I cannot do, unless you could step out of your Line altogether.

King. Out of my Line? Do you mean out of the world? Out of Space?

I. Well, yes. Out of *your* World. Out of *your* Space. For your Space is not the true Space. True Space is a Plane;[3] but your Space is only a Line.

King. If you cannot indicate this motion from left to right by yourself moving in it, then I beg you to describe it to me in words.

I. If you cannot tell your right side from your left, I fear that no words of mine can make my meaning clear to you. But surely you cannot be ignorant of so simple a distinction.

King. I do not in the least understand you.

I. Alas! How shall I make it clear? When you move straight on, does it not sometimes occur to you that you *could* move in some other way, turning your eye round so as to look in the direction towards which your side is now fronting? In other

3 A masterly development of the dimensional analogy: Just as the line of Lineland is merely part of the plane of Flatland, so A. Square is shortly going to be told that the plane of Flatland is merely part of the solid of Spaceland. And the Victorian reader is going to be told that the solid of Spaceland is merely part of … what?

words, instead of always moving in the direction of one of your extremities, do you never feel a desire to move in the direction, so to speak, of your side?

King. Never. And what do you mean? How can a man's inside "front" in any direction? Or how can a man move in the direction of his inside?

I. Well then, since words cannot explain the matter, I will try deeds, and will move gradually out of Lineland in the direction which I desire to indicate to you.

At the word I began to move my body out of Lineland. As long as any part of me remained in his dominion and in his view, the King kept exclaiming, "I see you, I see you still; you are not moving." But when I had at last moved myself out of his Line, he cried in his shrillest voice, "She is vanished; she is dead." "I am not dead," replied I; "I am simply out of Lineland, that is to say, out of the Straight Line which you call Space, and in the true Space, where I can see things as they are. And at this moment I can see your Line, or side — or inside as you are pleased to call it; and I can see also the Men and Women on the North and South of you, whom I will now enumerate, describing their order, their size, and the interval between each."

When I had done this at great length, I cried triumphantly, "Does that at last convince you?" And, with that, I once more entered Lineland, taking up the same position as before.

But the Monarch replied, "If you were a Man of sense—though, as you appear to have only one voice I have little doubt you are not a Man but a Woman—but, if you had a particle of sense, you would listen to reason. You ask me to believe that there is another Line[4] besides that which my senses indicate, and another motion besides that of which I am daily conscious. I, in return, ask you to describe in words or indicate by motion that other Line of which you speak. Instead of moving, you merely exercise some magic art of vanishing and returning to sight; and instead of any lucid description of your new World, you simply tell me the numbers and sizes of some forty of my retinue, facts known to any child in my capital. Can anything be more irrational or audacious? Acknowledge your folly or depart from my dominions."

Furious at his perversity, and especially indignant that he professed to be ignorant of my sex, I retorted in no measured terms, "Besotted Being! You think yourself the perfection of existence, while you are in reality the most imperfect and imbecile. You profess to see, whereas you can see nothing but a Point! You plume yourself on infer-

4 Here Abbott is thinking in terms of Cartesian coordinates, with Lineland as one axis of the Flatland plane. There must be *another* axis, perpendicular to the first. All the King of Lineland has to do is move out of his own universe, along the direction that is obvious to A. Square.

5 There are echoes here of the biblical phrase *king of kings,* with which Abbott would have been familiar. Just as a king of kings is superior to a mere king, so a line of lines (A. Square) is superior to a mere line, and Abbott may have used the phrase consciously or subconsciously with this in mind. It also has a direct mathematical meaning that is central to Abbott's explanations of his "dimensional analogy." A square can be considered as a family of parallel lines—all of the slices in the east-west direction, say. These slices stack along a second line, the north-south direction. Hence a square is a "line of lines." This allows Abbott to introduce a cube as a "line of squares," later in the tale.

In modern mathematics, Abbott's phrase becomes a literal description. The "real line" R corresponds to all real numbers x—the point labeled x lies distance x along the line relative to a fixed origin, 0. Positive x is measured eastward, negative x westward. A plane consists of all *pairs* (x, y) of real numbers: distances along two mutually perpendicular lines, the coordinate axes. Thus the plane is the "Cartesian product" $R \times R$, or R^2—a "line's worth" (first R) of lines (second R). See Figure 26.

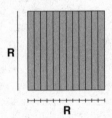

Figure 26 The plane as a line's worth of lines.

ring the existence of a Straight Line; but I *can see* Straight Lines, and infer the existence of Angles, Triangles, Squares, Pentagons, Hexagons, and even Circles. Why waste more words? Suffice it that I am the completion of your incomplete self. You are a Line, but I am a Line of Lines,[5] called in my country a Square: and even I, infinitely superior though I am to you, am of little account among the great nobles of Flatland, whence I have come to visit you, in the hope of enlightening your ignorance."

Hearing these words the King advanced towards me with a menacing cry as if to pierce me through the diagonal; and in that same moment there arose from myriads of his subjects a multitudinous war-cry, increasing in vehemence till at last methought it rivalled the roar of an army of a hundred thousand Isosceles, and the artillery of a thousand Pentagons. Spell-bound and motionless, I could neither speak nor move to avert the impending destruction; and still the noise grew louder, and the King came closer, when I awoke to find the breakfast-bell recalling me to the realities of Flatland.

§ 15. — *Concerning a Stranger from Spaceland.*

FROM DREAMS I proceed to facts.

It was the last day of the 1999th year of our era. The pattering of the rain had long ago announced nightfall; and I was sitting[1] in the company of my wife, musing on the events of the past and the prospects of the coming year, the coming century, the coming Millennium.

My four Sons and two orphan Grandchildren had retired to their several apartments; and my wife alone remained with me to see the old Millennium out and the new one in.

I was rapt in thought, pondering in my mind some words that had casually issued from the mouth of my youngest Grandson, a most promising young Hexagon of unusual brilliancy and perfect angularity. His uncles and I had been giv-

---

[1] When I say "sitting," of course I do not mean any change of attitude such as you in Spaceland[1] signify by that word; for as we have no feet, we can no more "sit" nor "stand" (in your sense of the word) than one of your soles or flounders.

Nevertheless, we perfectly well recognize the different mental states of volition implied in "lying," "sitting," and "standing," which are to some extent indicated to a beholder by a slight increase of lustre corresponding to the increase of volition.

But on this, and a thousand other kindred subjects, time forbids me to dwell.

1 The first edition has "... you in Flatland ...," an evident slip.

2 If three lines of length 1 inch (I use inches here because Abbott does and omit metric equivalents) are joined end to end, the result is a line of length 3 inches, the same shape as the original line but three times as large (Figure 27a). However, a square cannot be made three times as large by joining together three copies: Instead, it takes nine such copies (Figure 27b). Note that 9 is the square of 3 — that is, $9 = 3^2 = 3 \times 3$. And the power 2 here is the dimension of a square.

In "What Is the Fourth Dimension?" C. H. Hinton uses a similar argument, only there he uses powers of 2 rather than 3. "Casting Out the Self," from his first volume of *Scientific Romances,* has a long discussion of a 3 x 3 x 3 cube.

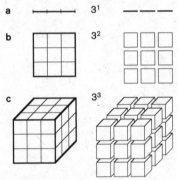

a |———————| 3^1

b 3^2

c 3^3

Figure 27 How figures of different dimensions scale: (**a**) line; (**b**) square; (**c**) cube.

3 Arithmetically, there is no difficulty in extending the notion of "square" to higher powers:

$3^3 = 3 \times 3 \times 3 = 27$ (cube)

$3^4 = 3 \times 3 \times 3 \times 3 = 81$ (fourth power)

$3^5 = 3 \times 3 \times 3 \times 3 \times 3 = 243$ (fifth power)

$3^6 = 3 \times 3 \times 3 \times 3 \times 3 \times 3 = 729$ (sixth power)

and so on. To *us* the cube has a clear geometric meaning (as the term suggests). To wit, if 27 unit cubes are stacked together in the right way (Figure 27c), then they form a cube three times the size. And here the power 3 is the dimension of the cube.

ing him his usual practical lesson in Sight Recognition, turning ourselves upon our centres, now rapidly, now more slowly, and questioning him as to our positions; and his answers had been so satisfactory that I had been induced to reward him by giving him a few hints on Arithmetic, as applied to Geometry.

Taking nine Squares,[2] each an inch every way, I had put them together so as to make one large Square, with a side of three inches, and I had hence proved to my little Grandson that — though it was impossible for us to *see* the inside of the Square — yet we might ascertain the number of square inches in a Square by simply squaring the number of inches in the side: "and thus," said I, "we know that 3^2, or 9, represents the number of square inches in a Square whose side is 3 inches long."

The little Hexagon meditated on this a while and then said to me; "But you have been teaching me to raise numbers to the third power:[3] I suppose 3^3 must mean something in Geometry; what does it mean?" "Nothing at all," replied I, "not at least in Geometry; for Geometry has only Two Dimensions." And then I began to shew the boy how a Point by moving through a length of three inches makes a Line of three inches, which may be represented by 3; and how a Line of three inches, moving parallel to itself through a length of three inches, makes a Square of three inches every way, which may be represented by 3^2.

Upon this, my Grandson, again returning to his former suggestion, took me up rather suddenly and exclaimed, "Well, then, if a Point by moving three inches, makes a Line of three inches represented by 3; and if a straight Line of three inches, moving parallel to itself, makes a Square of three inches every way, represented by 3^2; it must be that a Square of three inches every way, moving somehow parallel to itself (but I don't see how) must make Something else (but I don't see what) of three inches every way—and this must be represented by 3^3."

"Go to bed," said I, a little ruffled by this interruption: "if you would talk less nonsense, you would remember more sense."

So my Grandson had disappeared in disgrace; and there I sat by my Wife's side, endeavouring to form a retrospect of the year 1999 and of the possibilities of the year 2000, but not quite able to shake off the thoughts suggested by the prattle of my bright little Hexagon. Only a few sands now remained in the half-hour glass. Rousing myself from my reverie I turned the glass Northward for the last time in the old Millennium; and in the act, I exclaimed aloud, "The boy is a fool."

Straightway I became conscious of a Presence in the room, and a chilling breath thrilled through my very being. "He is no such thing," cried my Wife, "and you are breaking the Commandments in thus dishonouring your own Grandson." But I took no notice of her. Looking round in every di-

A. Square's bright little hexagonal grandson is convinced that because 3^3 has an algebraic meaning, it must also have a geometric one, but grandfather (at this stage of the narrative) disagrees. In the same way, a Victorian reader would disagree that 3^4 has a geometric meaning. The ancient Greek geometers had a similar attitude. They used not raw numbers but *magnitudes,* numbers with units — specifically, lengths of lines. The product of two numbers was thus an area, that of three was a volume, and that of four was … they didn't say. They rather slid over the problem and felt uncomfortable when more than three magnitudes were multiplied together. However, such a multiplication occurs in a famous (and useful) theorem, *Heron's Formula* for the area of a triangle. If the sides of the triangle are *a*, *b*, and *c*, and we define the semiperimeter *s = (a + b + c)/2,* then the area is the square root of *s(s - a)(s - b)(s - c).* The formula thus involves the product of four lengths. One way out of the dilemma (which seems not to have occurred to the Greeks) is to rewrite this as the square root of *s(s - a)* multiplied by the square root of *(s - b)(s - c).* Each of these products involves just two numbers, and the square roots reduce them from areas to lengths, so the theorem expresses the area of the triangle as a product of two lengths, with no interpretational difficulties.

Related ideas have been extended to more exotic shapes, such as the *Sierpinski gasket* (Figure 28)

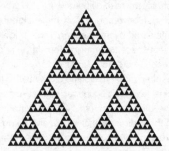

Figure 28 The Sierpinski gasket. The figure shows an early stage; the process of subdividing triangles should be continued indefinitely.

invented by the Polish mathematician Waclaw Sierpinski (1882–1969). If k copies of a shape can be joined together to make an exact copy that is b times as big, then the *similarity dimension* of the shape is

$$\log k \,/\, \log b$$

where log is the natural logarithm. For a cube, $k = 27$ and $b = 3$, and

$$\log 27 \,/\, \log 3 = 3$$

as expected. For the Sierpinski gasket, $k = 3$ and $b = 2$, and

$$\log 3 \,/\, \log 2 = 1.585$$

which is *not a whole number.* Similarity dimension, unlike "number of directions," need not take whole-number values. The Sierpinski gasket is an example of a *fractal,* a geometric shape with fine structure on all scales of magnification (see Benoit Mandelbrot, *The Fractal Geometry of Nature*; and chapter 5 of *Flatterland,* "One and a Quarter Dimensions"). The similarity dimension has a profound generalization, the *Hausdorff–Besicovitch* or *fractal* dimension (see Kenneth Falconer, *Fractal Geometry*).

4 When a sphere intersects a plane, the result is a circle. As the sphere moves relative to the plane, the circle changes size; observe that (page 143) Abbott draws the intersections as having sharp ends when actually they are ellipses — circles viewed in perspective. Conversely, a high-dimensional shape can be visualized by intersecting it with a lower-dimensional space and moving the shape relative to that space to create a series of "cross sections." In this manner, mathematicians can visualize four-dimensional forms as sequences of three-dimensional slices, five-dimensional forms as two-dimensional arrays of three-dimensional slices, and so on. This technique can be traced way back into mathematical folklore. One person who made it explicit, and placed great store by it, was C. H. Hinton. The story involves a curious and complicated series of influ-

rection I could see nothing; yet still I *felt* a Presence, and shivered as the cold whisper came again. I started up. "What is the matter?" said my Wife, "there is no draught; what are you looking for? There is nothing." There was nothing; and I resumed my seat, again exclaiming, "The boy is a fool, I say; 3^3 can have no meaning in Geometry." At once there came a distinctly audible reply, "The boy is not a fool; and 3^3 has an obvious Geometrical meaning."

My Wife as well as myself heard the words, although she did not understand their meaning, and both of us sprang forward in the direction of the sound. What was our horror when we saw before us a Figure! At the first glance it appeared to be a Woman, seen sideways; but a moment's observation shewed me that the extremities passed into dimness too rapidly to represent one of the Female Sex; and I should have thought it a Circle,[4] only that it seemed to change its size in a manner impossible for a Circle or for any regular Figure of which I had had experience.

But my Wife had not my experience, nor the coolness necessary to note these characteristics. With the usual hastiness and unreasoning jealousy of her Sex, she flew at once to the conclusion that a Woman had entered the house through some small aperture. "How comes this person here?" she exclaimed, "you promised me, my dear, that there should be no ventilators in our new house." "Nor are there any," said I; "but what makes you think that the stranger is a Woman? I see by my power of Sight Recognition——" "Oh, I have no patience with your Sight Recognition," replied she, "'Feeling is believing' and 'A Straight Line to the touch is worth a Circle to the sight'"—two Proverbs, very common with the Frailer Sex in Flatland.

"Well," said I, for I was afraid of irritating her, "if it must be so, demand an introduction." Assuming her most gracious manner, my Wife advanced towards the Stranger, "Permit me, Madam, to feel and be felt by——" then, suddenly recoiling, "Oh! it is not a Woman, and there are no angles either, not a trace of one. Can it be that I have so misbehaved to a perfect Circle?"

"I am indeed, in a certain sense a Circle," replied the Voice, "and a more perfect Circle than any in Flatland; but to speak more accurately, I am many Circles in one." Then he added more mildly, "I have a message, dear Madam, to your husband, which I must not deliver in your presence; and, if you would suffer us to retire for a few minutes——" But my Wife would not listen to the proposal that our august Visitor should so incommode himself, and assuring the Circle that the hour of her own retirement had long passed, with many reiterated apologies for her recent indiscretion, she at last retreated to her apartment.

I glanced at the half-hour glass. The last sands had fallen. The third Millennium had begun.[5]

ences and relationships that revolved around Alicia Stott (née Boole), third daughter of the English mathematician George Boole (1815–1864). There seems to be a strong historical link among Hinton's four-dimensional speculations, some brilliant mathematical discoveries of Alicia's own, and Abbott's writing of *Flatland*.

George Boole is best known for *Boolean algebra,* a form of symbolic logic in which 1 represents "true," 0 represents "false," and the logical operations *not, and,* and *or* are formulated algebraically. It is widely used today as a basis for computer science. It also gave a big boost to the development of mathematical logic and eventually led to deep and important work on the foundations of mathematics. This includes the epic theorems of Kurt Gödel (1906–1978), who proved that the logical consistency of arithmetic can never be proved and that some statements in arithmetic are undecidable— neither provable nor disprovable. Boole, the son of a Lincoln shoemaker, was a child prodigy, but his education came to an abrupt halt when he was sixteen because of the family's financial circumstances. His two books *The Mathematical Analysis of Logic* (1847) and *An Investigation of the Laws of Thought* (1854) laid the foundations of theoretical computer science. (Lewis Carroll was also interested in mathematical logic and published *Symbolic Logic I* in 1896. There is no second volume.) Boole also worked on the theory of invariants, differential equations, and probability. In 1849 he was appointed first professor of mathematics in Ireland's new Queen's College, now University College, Cork.

In 1855 he married Mary Everest (1832–1916), daughter of the Reverend Thomas Roupell Everest, Rector of Wickwar, Gloucestershire. Her uncle was Lieutenant Colonel Sir George Everest (1790–1866), after whom the world's highest mountain is named. Mary showed an early aptitude for mathematics, though in later life her knowledge of the subject seems poor. In 1852 Boole visited the Everest family and began giving Mary lessons in mathematics;

their relationship grew from there, although Boole considered himself too old to marry. His mind changed abruptly when Mary's father died in 1855, leaving her destitute: Boole immediately proposed. Mary was for a time secretary to James Hinton, a former friend of her father's, who was an ear surgeon, an enthusiastic advocate of sexual freedom, and C. H. Hinton's father. Mary promoted the elder Hinton's ideas; she was a prolific writer of extremely uneven quality, and her work ranges from the insightful to the bizarre. She also tried to promote her husband's investigations in logic, with dire results.

The Booles had five daughters: Mary Ellen (1856–?), Margaret (1858–?), Alicia (1860–1940), Lucy Everest (1862–1905), and Ethel Lilian (1864–1960). Mary Ellen married C. H. Hinton. Three of Hinton's books are important here: *A New Era of Thought* (1884), *The Fourth Dimension* (1904), and *An Episode of Flatland* (1907). In the first two of these, Hinton developed a way to visualize four-dimensional solids as a series of three-dimensional cross sections. He used a set of 81 ($= 3^4$) tiny wooden cubes, colored and labeled in a highly ingenious manner, to aid his thoughts. (Earlier, he had assigned a two-word Latin name to each of the 46,656 ($= 36^3$) 1-inch subcubes of a cube whose side was 1 yard — 36 inches.) According to McHale, "… it is quite probable that [Hinton's] interest in these topics was stimulated by his mother-in-law Mary Boole and her daughters." *An Episode of Flatland* is not set on Abbott's Flatland; it is Hinton's Flatlandesque tale of Astria. As we have already mentioned, Hinton wrote several articles about two-dimensional worlds in the early 1880s, and Banchoff has suggested that "… it is likely that Abbott could have seen one or more of [Hinton's] articles before he wrote *Flatland*." Another connection should be noted: Mary Boole (senior) was a friend of Herbert George Wells (1866–1946), and there is a fictional character named Boole in his *The New Machiavelli* (1911). Wells's *The Time Machine* (1895) takes its scientific inspiration from the geometry of the fourth dimension, and it seems that Wells was influenced in this by Hinton.

The second Boole daughter, Margaret, married the artist Edward Ingram Taylor. Their son Geoffrey Ingram Taylor became one of the leading applied mathematicians of the twentieth century: He worked on (among other topics) fluid mechanics, shock waves, meteorology, icebergs, crystals, and aeronautics.

It is the third daughter, Alicia (generally known as Alice), who provides the strongest mathematical link to *Flatland* — not by influencing its writing but by developing Hinton's ideas, upon which it seems clear that *Flatland* was based, into serious mathematics. Alicia possessed remarkable innate mathematical talents (her father died when she was four, so his possible influence as a teacher can be ruled out). At a time when women were actively discouraged from taking up mathematics, Alice studied Euclid. At the age of eighteen she became entranced by C. H. Hinton's tiny wooden cubes. She failed to share his fascination with the mystical aspects of the fourth dimension but found its *geometry* fascinating. She developed Hinton's methods (series of three-dimensional sections) and introduced the name *polytope* for a multidimensional solid. By purely geometric reasoning, she discovered that there are exactly six regular four-dimensional polytopes, formed from 5, 16, or 600 tetrahedra, 8 cubes, 24 octahedra, or 120 dodecahedra. She built beautiful cardboard models of their three-dimensional sections but then abandoned mathematics in 1890 to marry Walter Stott. The story might have ended there, but around 1900 she came across a paper by Peiter Hendrik Schoute (1846–1923) of Groningen University, who had obtained exactly the same results by purely analytic methods. Schoute encouraged her to publish her work, which she did in 1900 and 1910, and they collaborated until his death in 1913. In 1930 her nephew G. I. Taylor introduced her to the geometer H(arold) S(cott) M(acdonald) Coxeter (1907–), who rekindled her interest in four-dimensional polytopes and described her work in his *Regular Polytopes* (1948, 2nd ed. 1963).

Lucy, the fourth daughter, did not marry and died young. She learned chemistry to get a job in a chemist's shop and went on to become a fellow of the Institute of Chemistry and (it is believed) the first woman professor of chemistry at the Royal Free Hospital, London.

Perhaps the most remarkable of the Boole daughters was the last: Ethel Lilian (known as Lily in her youth). Around 1885 she met Sergei Kravchinski (pseudonym Stepniak), who had assassinated General N. V. Mezentsov. The general was Gendarme Chief and 3rd Section Head Controller; the 3rd Section was the Tsarist Secret Police. Kravchinski stabbed him in the stomach on a busy street, leaped into a passing *droshky* (open horse-drawn carriage), and fled Russia. Under Kravchinski's influence, Ethel Lilian became a political activist for the Society of Friends of Russian Freedom, a Marxist organization. In 1891 she married a Lithuanian refugee, Wilfred Michail Voynich (1865–1930) and embarked on a career as a writer. Her first book, *Stories from Garshin* (1893), was a translation of a book by Stepniak; she wrote many others, including the novel *The Gadfly* (1897), which is about a devout young man who loses his faith and is maimed, mutilated, and tortured to death. It was a wild success, especially in Russia, and other novels followed. In 1912 her husband bought a curious document, the "Voynich manuscript," which was written in code — some say by Roger Bacon (c.1220–1292). In 1960 it was sold for $160,000, and it is now in the vaults of Yale University. Despite intense efforts, it has not been deciphered. Cryptographic expert David Kahn called it "the most mysterious manuscript in the world."

The important aspect of this strange web of family influences is that it connects *Flatland,* albeit indirectly, to the mainstream mathematics of Abbott's day by way of a seminal figure in English mathematics: George Boole. It gives us a feeling for the Victorian social and scientific undercurrents out of which

Flatland emerged. The lives and work of Boole's five astonishing daughters add grace notes to this curious web of interconnections. And we will shortly see that Wells, a family friend of the Booles, is linked to Hinton and Abbott in another way, through his interest in the fourth dimension. The story of Boole, his family, and its influences is told in graphic detail by Desmond MacHale in his superb biography *George Boole, His Life and Work* (1985).

5 The first edition of *Flatland* has "second Millennium," which is wrong. So does at least one modern edition, the Shambhala one.

*§ 16. — How the Stranger vainly endeavoured
to reveal to me in words the mysteries
of Spaceland.*

As SOON AS the sound of the Peace-cry of my departing Wife had died away, I began to approach the Stranger with the intention of taking a nearer view and of bidding him be seated: but his appearance struck me dumb and motionless with astonishment. Without the slightest symptoms of angularity he nevertheless varied every instant with gradations of size and brightness scarcely possible for any Figure within the scope of my experience. The thought flashed across me that I might have before me a burglar or cut-throat, some monstrous Irregular Isosceles, who, by feigning the voice of a Circle, had obtained admission somehow into the house, and was now preparing to stab me with his acute angle.

In a sitting-room, the absence of Fog (and the season happened to be remarkably dry), made it difficult for me to trust to Sight Recognition, especially at the short distance at which I was standing. Desperate with fear, I rushed forward with an unceremonious, "You must permit me, Sir—" and felt him. My Wife was right. There was not the trace of an angle, not the slightest roughness or inequality: never in my life had I met with a more perfect Circle. He remained motionless while I walked round him, beginning from his eye and returning to it again. Circular he was throughout, a perfectly satisfactory Circle; there could not be a doubt of it. Then followed a dialogue, which I will endeavour to set down as near as I can recollect it, omitting only some of my profuse apologies—for I was covered with shame and humiliation that I, a Square, should have been guilty of the impertinence of feeling a Circle. It was commenced by the Stranger with some impatience at the lengthiness of my introductory process.

Stranger. Have you felt me enough by this time? Are you not introduced to me yet?

I. Most illustrious Sir, excuse my awkwardness, which arises not from ignorance of the usages of polite society, but from a little surprise and nervousness, consequent on this somewhat unexpected visit. And I beseech you to reveal my indiscretion to no one, and especially not to my Wife. But before your Lordship enters into further communications, would he deign to satisfy the curiosity of one who would gladly know whence his Visitor came?

Stranger. From Space, from Space, Sir: whence else?

I. Pardon me, my Lord, but is not your Lordship already in Space, your Lordship and his humble servant, even at this moment?

1 A. Square is now experiencing the same difficulty that the King of Lineland faced in his dream.

Stranger. Pooh! what do you know of Space? Define Space.

I. Space, my Lord, is height and breadth indefinitely prolonged.

Stranger. Exactly: you see you do not even know what Space is. You think it is of Two Dimensions only; but I have come to announce to you a Third —height, breadth, and length.

I. Your Lordship is pleased to be merry. We also speak of length and height, or breadth and thickness, thus denoting Two Dimensions by four names.

Stranger. But I mean not only three names, but Three Dimensions.

I. Would your Lordship indicate or explain to me in what direction is the Third Dimension,[1] unknown to me?

Stranger. I came from it. It is up above and down below.

I. My Lord means seemingly that it is Northward and Southward.

Stranger. I mean nothing of the kind. I mean a direction in which you cannot look, because you have no eye in your side.

I. Pardon me, my Lord, a moment's inspection will convince your Lordship that I have a perfect luminary at the juncture of two of my sides.

Stranger. Yes: but in order to see into Space you ought to have an eye, not on your Perimeter, but on your side, that is, on what you would probably call your inside; but we in Spaceland should call it your side.

I. An eye in my inside! An eye in my stomach![2] Your Lordship jests.

Stranger. I am in no jesting humour. I tell you that I come from Space, or, since you will not understand what Space means, from the Land of Three Dimensions whence I but lately looked down upon your Plane which you call Space forsooth. From that position of advantage I discerned all that you speak of as *solid* (by which you mean "enclosed on four sides"), your houses, your churches, your very chests and safes, yes even your insides and stomachs, all lying open and exposed to my view.

I. Such assertions are easily made, my Lord.

Stranger. But not easily proved, you mean. But I mean to prove mine.

When I descended here, I saw your four Sons, the Pentagons, each in his apartment, and your two Grandsons the Hexagons; I saw your youngest Hexagon remain a while with you and then retire to his room, leaving you and your Wife alone. I saw your Isosceles servants, three in number, in the kitchen at supper, and the little Page in the scullery. Then I came here, and how do you think I came?

I. Through the roof, I suppose.

Stranger. Not so. Your roof, as you know very well, has been recently repaired, and has no aperture by which even a Woman could penetrate. I tell you I come from Space. Are you not convinced by what I have told you of your children and household?

2 Abbott has a lot of fun with a curious feature of dimensionality. To the creatures of Flatland, confined to their plane, the interior of a Polygon is not directly visible because its edges get in the way. If the plane is viewed from a third dimension, the entire interior, including internal organs, becomes visible. However, A. Square does not just need an eye in his stomach: It has to be pointing *out of his universe.* It would be equally effective (and equally impossible) for him to point his existing eye in that direction. By analogy, if our three-dimensional world could be viewed by a creature from a fourth dimension, *our* own insides would be completely visible to that creature.

The bizarre phenomena that could easily be exploited by a being from the fourth dimension have long intrigued philosophers. Rudy Rucker refers to these collectively as hyperspace philosophers — the word hyperspace is often used for four-dimensional space (sometimes higher). One of the earliest hyperspace philosophers was Henry More (1614–1687), who disliked the idea of angels and spirits as disembodied, insubstantial beings; he argued that if they actually existed, they must take up space. However, there is a problem in fitting a spatially extended soul into a physical body: Where do you put it, and why hasn't medical science located it? All is solved if the spirit world is four-dimensional, a concept that, in his *Enchiridion Metaphysicum* of 1671, More called spissitude — a kind of "Extra-Height," in A. Square's terminology, or a slight thickening along a fourth dimension. Living bodies have more spissitude than dead ones. In 1747 Immanuel Kant (1724–1804) wrote,

> A science of all these possible kinds of space [with more than three dimensions] would undoubtably be the highest enterprise which a finite understanding could undertake in the field of geometry....If it is possible that there are extensions with other dimensions, it is also very probable that God has somewhere brought them into being: for His works have all the magnitude and manifoldness of which they are capable.

Cranks and eccentrics also found the concept of a fourth dimension fascinating, invoking it to explain the "spirit world" of the spiritualist movement. Four-dimensional space provides a place for the dead to continue existing, alongside our universe (and hence able to communicate with it) but not *in* our universe. A prominent advocate of this kind of mysticism was Peter Demianovich Ouspensky (1878–1947), who revealed the mysteries of the fourth dimension to the intellectuals of Tsarist Russia. Ouspensky was a disciple of George Ivanovitch Gurdjieff (1872–1949), once described in *Time* magazine as "a remarkable blend of P.T. Barnum, Rasputin, Freud, Groucho Marx, and everybody's grandfather." In his 1908 *The Fourth Dimension,* Ouspensky writes,

> If the fourth dimension exists while we possess only three, it means that we have no real existence, that we exist only in somebody's imagination and that all our thoughts, feelings and experiences take place in the mind of some other higher being ... the whole of our universe is but an artificial world created by his fantasy.
>
> If we do not want to agree with this we must recognize ourselves as beings in four dimensions.... [W]e have no need to think that the spirits that appear or fail to appear at spiritualistic seances must be the only possible beings of four dimensions. We may have very good reason for saying that we are ourselves beings of four dimensions and we are turned towards the third dimension with only one of our sides, i.e., with only a small part of our being.

His *Tertium Organum* of 1912 adds,

> When we shall see or feel ourselves in the world of four dimensions we shall see that the world of three dimensions does not really exist and has never existed; that it was the creation of our own fantasy, a phantom host, an optical illusion, a delusion — anything one pleases excepting only reality.

Ouspensky's influence shows up, very briefly, in Book 5, Section 3 of the 1879–1880 novel *The*

I. Your Lordship must be aware that such facts touching the belongings of his humble servant might be easily ascertained by any one in the neighbourhood possessing your Lordship's ample means of obtaining information.

Stranger. (*To himself.*) What must I do?[3] Stay; one more argument suggests itself to me. When you see a Straight Line — your wife, for example — how many Dimensions do you attribute to her?

I. Your Lordship would treat me as if I were one of the vulgar who, being ignorant of Mathematics, suppose that a Woman is really a Straight Line, and only of One Dimension. No, no, my Lord; we Squares are better advised, and are as well aware as your Lordship that a Woman, though popularly called a Straight Line, is, really and scientifically, a very thin Parallelogram,[4] possessing Two Dimensions, like the rest of us, viz., length and breadth (or thickness).

Stranger. But the very fact that a Line is visible implies that it possesses yet another Dimension.

I. My Lord, I have just acknowledged that a Woman is broad as well as long. We see her length, we infer her breadth; which, though very slight, is capable of measurement.

Stranger. You do not understand me. I mean that when you see a Woman, you ought — besides inferring her breadth — to see her length, and to *see* what we call her *height*; although that last Dimension is infinitesimal in your country.[5] If a Line were mere length without "height," it would cease to occupy Space and would become invisible. Surely you must recognize this?

I. I must indeed confess that I do not in the least understand your Lordship. When we in Flatland see a Line, we see length and *brightness.* If the brightness disappears, the Line is extinguished, and, as you say, ceases to occupy Space. But am I to suppose that your Lordship gives to brightness the title of a Dimension,[6] and that what we call "bright" you call "high"?

Stranger. No, indeed. By "height" I mean a Dimension like your length: only, with you, "height" is not so easily perceptible, being extremely small.[7]

I. My Lord, your assertion is easily put to the test. You say I have a Third Dimension, which you call "height." Now, Dimension implies direction and measurement. Do but measure my "height," or merely indicate to me the direction in which my "height" extends, and I will become your convert. Otherwise, your Lordship's own understanding must hold me excused.

Stranger. (*To himself.*) I can do neither. How shall I convince him? Surely a plain statement of facts followed by ocular demonstration ought to suffice. —Now, Sir; listen to me.

You are living on a Plane. What you style Flatland is the vast level surface of what I may call a fluid,[8] on, or in, the top of which you and your countrymen move about, without rising above it or falling below it.

I am not a plane Figure, but a Solid. You call

Brothers Karamazov by Fyodor Dostoyevsky (1821–1881), in a passage where Ivan Karamazov pursues possible connections between high-dimensional and non-Euclidean geometries and the existence of God:

> ... if God exists and if he really did create the earth then, as common knowledge tells us, he created it according to Euclidean geometry, while he created the human mind with an awareness of only three spatial dimensions. Even so, there have been and still are even today geometers and philosophers of the most remarkable kind who doubt that the entire universe or, even more broadly, the entirety of being was created solely according to Euclidean geometry.

Vladimir Ilich Lenin (1870–1924) was sufficiently worried by this sort of thing that in 1908 he wrote *Materialism* and *Empirio-Criticism,* attacking the idea of spirits in four-dimensional space. Mathematics, said Lenin, may open up the possibility of space of four dimensions, but only in three-dimensional space can political revolution occur.

3 This begins one of the two new passages that Abbott mentions in his preface to the second edition. The insertion continues until "*Stranger.* (*To Himself.*) I can do neither?" on page 141. In place of this new passage, the first edition has: "*Stranger.* How shall I convince him?"

4 Abbott is setting the reader up for the Stranger's subsequent argument that everything in Flatland *really* possesses a third dimension but is very "thin" in that direction. Otherwise, the Stranger argues, Women would be invisible in Flatland. Abbott's reasoning here seems confused. In a three-dimensional world — space — light is confined to the space and is scattered by surfaces. The brightness of an object is determined by how much light it scatters per unit area. Analogously, in a two-dimensional world — a plane — light is confined to the plane and is scattered by lines and curves. The brightness of an object is determined by how much light it scatters per unit *length*. The edge of a polygon is a line and possesses nonzero length, hence it can scatter light, hence it can be visible. But the same goes for a Woman, who is a line and hence also possesses nonzero length. Only when the Woman is viewed end-on would she become invisible — and Abbott has already told us as much.

There is another possibility, though. Thin objects generally allow more light to pass *through* them than thick ones do. Flatland's Women might, perhaps, be transparent. It seems clear that this is not what Abbott has in mind, but even if he did, a very thin parallelogram would also be virtually transparent. So would a very narrow isosceles triangle, especially near the sharp end.

It therefore seems necessary to assume that Women are opaque. A. Square would then have no more difficulty in seeing a Woman than he would in seeing anything else, and her (near) lack of a second dimension would be irrelevant. No wonder A. Square is not convinced by the Stranger's argument.

Although A. Square can readily see Women, the *Stranger* will have problems. His visual senses are those of a three-dimensional world, and to him, brightness is determined by the amount of light reflected per unit *area*. A Woman, with (near) zero area, would not be visible to the Stranger, but a Polygon, with nonzero area, would be — except when viewed edge-on.

5 Abbott is using *infinitesimal* in the sense of "very small," not in the technical sense of "infinitely small" (which requires some extensive contortions in mathematical logic before it can be endowed with a sensible meaning that is different from plain zero [see Ian Stewart, *From Here to Infinity*]). But again, there is a problem with the Stranger's argument. If the height of an object is extremely small, then the extra amount of light that it scatters by virtue of its height is also extremely small. Therefore, height does not solve the visibility problem (for the Stranger, that is; there is no problem for A. Square in any case).

me a Circle; but in reality I am not a Circle, but an infinite number of Circles, of size varying from a Point to a Circle of thirteen inches in diameter, one placed on the top of the other. When I cut through your plane as I am now doing, I make in your plane a section which you, very rightly, call a Circle. For even a Sphere—which is my proper name in my own country—if he manifest himself at all to an inhabitant of Flatland—must needs manifest himself as a Circle.

Do you not remember—for I, who see all things, discerned last night the phantasmal vision of Lineland written upon your brain—do you not remember, I say, how, when you entered the realm of Lineland, you were compelled to manifest yourself to the King, not as a Square, but as a Line, because that Linear Realm had not Dimensions enough to represent the whole of you, but only a slice or section of you? In precisely the same way, your country of Two Dimensions is not spacious enough to represent me, a being of Three, but can only exhibit a slice or section[9] of me, which is what you call a Circle.

The diminished brightness of your eye indicates incredulity. But now prepare to receive proof positive of the truth of my assertions. You cannot indeed see more than one of my sections, or Circles, at a time; for you have no power to raise your eye out of the plane of Flatland; but you can at least see that, as I rise in Space, so my sections become smaller. See now, I will rise; and the effect

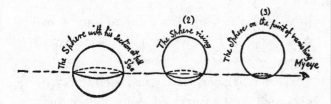

upon your eye will be that my Circle will become smaller and smaller till it dwindles to a point and finally vanishes.

There was no "rising" that I could see;[10] but he diminished and finally vanished. I winked once or twice to make sure that I was not dreaming. But it was no dream. For from the depths of nowhere came forth a hollow voice[11] — close to my heart it seemed — "Am I quite gone? Are you convinced now? Well, now I will gradually return to Flatland and you shall see my section become larger and larger."

Every reader in Spaceland will easily understand that my mysterious Guest was speaking the language of truth and even of simplicity. But to me, proficient though I was in Flatland Mathematics, it was by no means a simple matter. The rough diagram given above will make it clear to any Spaceland child that the Sphere, ascending in the three positions indicated there, must needs have manifested himself to me, or to any Flatlander, as a Circle, at first of full size, then small, and at last very small indeed, approaching to a Point. But to me, although I saw the facts before me, the causes were as dark as ever. All that I could comprehend was, that the Circle had made

If the Stranger's argument were correct, then by Abbott's central analogy, a similar argument would apply in three dimensions. That is, in order for an object to be visible, it would have to possess some kind of "thickness" along a *fourth* dimension. However, the physics of light in ordinary space — which Abbott would have known about, being keen on the teaching of science as well as classics — requires no such assumption. The Stranger's argument is paralleled in the preface to the second edition of *Flatland,* with its reference to "Extra-Height," but Abbott uses this idea to *demolish* the argument that Flatlanders would not be able to see other Flatlanders unless they all extended slightly into a third dimension.

It rather looks as though Abbott's critics (of the first edition) agreed with the Stranger and threw the argument back at him; Abbott responded correctly in the preface but failed to change the section, including the Stranger's fallacious argument.

6 The Stranger objects to this suggestion: He is referring to a *spatial* dimension. The modern view is that brightness, being an independent quantity, *could* be considered an extra dimension — but not in the Stranger's sense.

7 Having just said that the physics of light requires no additional small "hidden" dimension, I must point out that this is the case only in Newtonian physics and in today's quantum physics. Tomorrow's physics may change that statement. Or yesterday's: In 1919 Theodor Kaluza (1885–1954) sent Einstein a proposal to unify gravity and electromagnetism by introducing a *fifth* dimension into space-time. Why don't we notice the extra dimension? Because it is curled up very, very tightly and it is very, very small. This is precisely Hinton's explanation of why the fourth dimension isn't obvious, for in "What Is the Fourth Dimension?" he says,

> The other alternative [to our being merely three-dimensional, if a fourth dimension exists] is that we

have a four-dimensional existence. In this case our proportions in it must be infinitely minute, or we should be conscious of them. If such be the case, it would probably be in the ultimate particles of matter, that we should discover the fourth dimension.

The final remark was oddly prophetic, as we shall see in a moment.

Kaluza's theory was improved by Oskar Klein (1894–1977) in 1926, but physicists lost interest in the Kaluza-Klein theory because it was extremely difficult to test experimentally. The idea of introducing extra dimensions into physics lived on, however. In recent attempts to devise a Theory of Everything (a unification of relativity and quantum theory), physicists have been led to postulate the existence of *superstrings* — particles that are not points but tiny multidimensional surfaces. Currently, the most popular version of this theory requires space-time to have its four usual dimensions *plus* six more, the extra six again being curled up into such a tiny region that they are imperceptible to any measuring instruments. The size of this curled-up region is expected to be roughly 1 Planck length — about 10^{-35} cm, the smallest distance that can be observed in a quantum universe. This idea is curiously reminiscent of Hinton's remark about "ultimate particles of matter." At any rate, physicists now think that our space-time might actually be ten-dimensional. See chapter 16 of *Flatterland*, "No-branes and *p*-branes."

8 Abbott's image here is mathematically and scientifically unnecessary, but he needed something tangible to carry his readers with him. Describing Flatland as the surface of a fluid has two narrative advantages. It makes it plausible that the inhabitants of Flatland can move at will across its planar surface: They are like thin wooden shapes floating on a pond. And it allows the Stranger to move *through* Flatland, like a fish rising from the deeps and breaking the surface of that pond. If Flatland were the surface of a solid, or just a rigid plane in its own right, this free-

himself smaller and vanished, and that he had now reappeared and was rapidly making himself larger.

When he regained his original size, he heaved a deep sigh; for he perceived by my silence that I had altogether failed to comprehend him.[12] And indeed I was now inclining to the belief that he must be no Circle at all, but some extremely clever juggler; or else that the old wives' tales were true, and that after all there were such people as Enchanters and Magicians.

After a long pause he muttered to himself, "One resource alone remains, if I am not to resort to action. I must try the method of Analogy." Then followed a still longer silence, after which he continued our dialogue.

Sphere. Tell me, Mr. Mathematician;[13] if a Point moves Northward, and leaves a luminous wake, what name would you give to the wake?

I. A straight Line.

Sphere. And a straight Line has how many extremities?

I. Two.

Sphere. Now conceive the Northward straight Line moving parallel to itself, East and West, so that every point in it leaves behind it the wake of a straight Line. What name will you give to the Figure thereby formed? We will suppose that it moves through a distance equal to the original straight Line. — What name, I say?

I. A Square.

Sphere. And how many sides has a Square? How many angles?

I. Four sides and four angles.

Sphere. Now stretch your imagination a little, and conceive a Square in Flatland, moving parallel to itself upward.

I. What? Northward?

Sphere. No, not Northward; upward; out of Flatland altogether.

If it moved Northward, the Southern points in the Square would have to move through the positions previously occupied by the Northern points. But that is not my meaning.

I mean that every Point in you—for you are a Square and will serve the purpose of my illustration—every Point in you, that is to say in what you call your inside, is to pass upwards through Space in such a way that no Point shall pass through the position previously occupied by any other Point; but each Point shall describe a straight Line of its own. This is all in accordance with Analogy; surely it must be clear to you.

Restraining my impatience—for I was now under a strong temptation to rush blindly at my Visitor and to precipitate him into Space, or out of Flatland, anywhere, so that I could get rid of him—I replied:—

"And what may be the nature of the Figure which I am to shape out by this motion which you are pleased to denote by the word 'upward'? I presume it is describable in the language of Flatland."

dom of movement would be harder to explain. It is also possible that Abbott had in mind the "luminiferous ether," a hypothetical fluid filling the whole of space, which Victorian scientists believed was the medium in which waves of electromagnetic radiation moved.

With suitable laws of physics, though, there is no need for familiar images of this kind. The plane of Flatland could be just a part of some larger space, and the physical laws could confine Flatlanders to this plane, while permitting other things (such as the Stranger) to exist in the larger space. In fact, current physical speculations place human beings in an analogous position. In "standard" superstring theory, our universe has ten dimensions, six of them being curled up so tightly that they cannot be perceived; human beings extend into those six extra dimensions, but they don't notice them. However, there is a recent variant in which the extra dimensions required for superstrings can be imperceptible without being curled up into a tiny ball. Instead, our universe (and we with it) might be confined to a four-dimensional "surface" in a ten-dimensional space-time, just as Flatland is confined to a two-dimensional surface of a three-dimensional space. The other six dimensions can then be very big; we just can't perceive them because our senses don't point in those directions.

9 The Stranger (now revealed as a Sphere) can exhibit his shape to A. Square only through his intersection with the plane of Flatland because this is all that A. Square can perceive. The intersection is a circle, whose radius depends on the position of the center of the Sphere relative to that plane. As this position changes, so does the circle's size. It is at its greatest when the Sphere's center lies on the plane, and the intersection is then the "equator" of the Sphere and has the same diameter as the Sphere itself. It is at its smallest when the Sphere is tangent to the plane, at which stage it forms a single point—a circle of zero diameter.

10 A Spaceland observer sees a rigid body, the Sphere, rising. A. Square's more limited perceptions enable him to see only a circle, which *changes*: Its size varies with *time*. It is important to understand that Abbott is *not* trying to develop the relativistic concept of time as a fourth dimension (or, in his analogy, of time as a third dimension augmenting Flatland's two spatial ones). Imagine that A. Square could make a movie of the changing circle — a series of still frames showing the sequence of circular intersections created by the Sphere. Then the geometry of the Sphere is encoded, in a manner that is entirely adequate for mathematical purposes, in the sequence of still frames. It is not necessary to *run* the movie and turn those frames into motion over time. By moving through the plane of Flatland, the Sphere *converts* its third spatial dimension into something that a Flatlander can identify with the passage of time. But time itself is very different from the Sphere's third spatial dimension, which exists at every instant of time.

11 This passage bears a distinct resemblance to act 1, scene 5 of *Hamlet,* where the ghost of Hamlet's father can be heard speaking from beneath the stage, immediately prior to the line quoted by Abbott at the top of the cover of *Flatland.*

12 We sympathize with A. Square, for we also have difficulty comprehending four spatial dimensions. Our brains have evolved in a (spatially) three-dimensional world, and the nerve cells in the parts of the brain that detect and analyze visual perception have evolved an architecture that allows them to structure the incoming signals so that they provide an adequate internal *representation* of that external world. Here I am not merely saying that the brain is a three-dimensional object, although in a literal sense it is. The brain is a network composed of a large but finite number of nerve cells linked together by linear connections, so for most purposes the brain can be considered to be intricate but one-

Sphere. Oh, certainly. It is all plain and simple, and in strict accordance with Analogy — only, by the way, you must not speak of the result as being a Figure, but as a Solid. But I will describe it to you. Or rather not I, but Analogy.

We began with a single Point, which of course — being itself a Point — has only *one* terminal Point.

One Point produces a Line with *two* terminal Points.

One Line produces a Square with *four* terminal Points.

Now you can give yourself the answer to your own question: 1, 2, 4, are evidently in Geometrical Progression. What is the next number?

I. Eight.

Sphere. Exactly. The one Square produces a *Something-which-you-do-not-as-yet-know-a-name-for-but which-we-call-a-Cube* with *eight* terminal Points. Now are you convinced?

I. And has this Creature sides, as well as angles or what you call "terminal Points"?

Sphere. Of course; and all according to Analogy. But, by the way, not what *you* call sides, but what *we* call sides. You would call them *solids.*

I. And how many solids or sides will appertain to this Being whom I am to generate by the motion of my inside in an "upward" direction, and whom you call a Cube?

Sphere. How can you ask? And you a mathematician! The side of anything is always, if I may so say, one Dimension behind the thing. Consequently, as there is no Dimension behind a Point, a Point has 0 sides; a Line, if I may so say, has 2 sides (for the Points of a Line may be called by courtesy, its sides); a Square has 4 sides; 0, 2, 4; what Progression do you call that?

I. Arithmetical.

Sphere. And what is the next number?

I. Six.

Sphere. Exactly. Then you see you have answered your own question. The Cube which you will generate will be bounded by six sides, that is to say, six of your insides. You see it all now, eh?

"Monster," I shrieked, "be thou juggler, enchanter, dream, or devil, no more will I endure thy mockeries. Either thou or I must perish." And saying these words I precipitated myself upon him.

dimensional (as is any mathematical network). What matters in a network is not its dimension but its *connectivity*—how many connections there are and what connects to what. A network that is connected like a grid in three-dimensional space is in effect three-dimensional, even if it is realized in the plane, and the same goes for four, five, or more dimensions.

In principle, then, a brain in three-dimensional space could evolve a neural network whose architecture *mimics* that of four-dimensional space, just as a brain in two-dimensional space could evolve a neural network whose architecture mimics that of three-dimensional space. However, such a structure would not evolve under natural conditions because there would be no systematic four-dimensional stimuli to cause it to do so. That having been said, such features as color and brightness in effect provide extra "dimensions" of stimuli, and our own visual systems *are* wired up to mimic these additional dimensions (so in a meaningful sense, we have six-dimensional circuitry inside our three-dimensional brains). But the brain interprets these extra dimensions as the "qualia" of color and brightness—the vivid sensory impressions with which the brain decorates them—and not as three more spatial dimensions.

13 We first learn that A. Square has mathematical leanings on p. 113, where we are told that he "... amused myself ... with my favourite recreation of Geometry." Abbott does not tell us A. Square's profession, and the Sphere's comment does not imply that he is a professional mathematician, only that the Sphere has noticed that he is mathematically inclined.

The following two and a half pages of Abbott's text provide a systematic discussion of the combinatorics of *n*-dimensional "cubes" for *n* = 1, 2, 3, which can be summarized in the table at the top of page 148. Here *vertex* means the same as *corner* (or *extremity* in Abbott's terminology), and I have used *face* to mean the (n - 1)-dimensional compo-

Dimension	Name	Number of Vertices	Number of Faces
0	point	1	0
1	line	2	2
2	square	4	4
3	cube	8	6

nents of an *n*-dimensional object: endpoints of a line, edges of a square, faces of a cube.

Note that Abbott recognizes a point as a 0-dimensional object. He has no hang-ups about the number 0. For instance, the Flatland calendar has a year 0, making it arguably more sensible than our own.

A. Square (who knows only the first three lines of the table from direct experience) is set two puzzles of the type "guess what number comes next." The first puzzle is

$$1, 2, 4, ?$$

and A. Square, being a mathematician, recognizes this as a *geometric sequence* (which in Abbott's day was called a geometrical progression). In a geometric sequence, every term bears a constant ratio to the previous one; in this case, each term is *twice* the previous one. Thus A. Square instantly knows that the next term must be 8.

The second puzzle is

$$0, 2, 4, ?$$

and A. Square recognizes this as an *arithmetic sequence* (which in Abbott's day was called an arithmetical progression). In an arithmetic sequence, every term has a constant difference from the previous one; in this case, each term is *two more than* the previous one. Thus A. Square instantly knows that the next term must be 6.

Abbott's argument here is flimsy but interesting. The dimensions accessible to A. Square show simple numerical patterns. Moreover, the Spaceland reader (and the Sphere) know that these patterns continue to three dimensions. The patterns are all the more compelling in that they embody the two

standard patterns taught in schools, both then and now. There are plenty of other types of sequences: *The Encyclopaedia of Integer Sequences* by Neil Sloane and Simon Plouffe lists 5,488 sequences of whole numbers, each obeying some specified rule of formation. One of the things that this book makes very clear is that "obvious" patterns need not continue. For example, what is the next term in the sequence 1, 2, 3, 4, 5? It could be 6 — if the rule is "the whole numbers." However, if the rule is "sums of two squares or three times a square," the next number is 8. And if the sequence is the "quadruple factorials" (defined recursively by $n!!!! = n(n-4)!!!!$), the next number is 12. Abbott's sequence 0, 2, 4 would continue with 8 if it were the sequence of integer parts $[n^2/2]$ of squares 1, 4, 9, 16, … divided by 2. And 1, 2, 4 would continue 8, 16, 31 (*not* 32) if the rule of formation were "join *n* points on a circle in all possible ways, and then count how many pieces the circle is sliced into" (Figure 29).

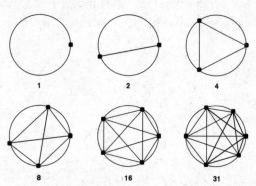

Figure 29 Into how many pieces do lines joining *n* points on a circle cut the circle? The early pattern does not continue in the "obvious" way.

§ 17. — How the Sphere, having in vain tried words, resorted to deeds.

IT WAS in vain. I brought my hardest right angle into violent collision with the Stranger, pressing on him with a force sufficient to have destroyed any ordinary Circle: but I could feel him slowly and unarrestably slipping from my contact; not edging to the right nor to the left, but moving somehow out of the world, and vanishing to nothing. Soon there was a blank. But still I heard the Intruder's voice.

Sphere. Why will you refuse to listen to reason? I had hoped to find in you — as being a man of sense and an accomplished mathematician — a fit apostle for the Gospel of the Three Dimensions,[1] which I am allowed to preach once only in a thousand years: but now I know not how to convince you. Stay, I have it. Deeds, and not words, shall proclaim the truth. Listen, my friend.

I have told you I can see from my position in Space the inside of all things that you consider closed. For example, I see in yonder cupboard near which you are standing, several of what you call boxes (but like everything else in Flatland, they have no tops nor bottoms) full of money; I

1 Abbott was ordained a deacon in the Anglican Church in 1862 and a priest in 1863, and most of his books are about religious matters. His use of *Gospel* here and the general fun he pokes at priests show that although he was serious about his religion, he was not solemn. The same goes for his science — how else could he have written a book like *Flatland*? Like all great educators, he combined a burning enthusiasm for his subject(s) with an aversion to pomposity.

2 This passage is a variant on the "I can see your insides" aspect of an extra dimension. In the same way, a creature from the fourth dimension could steal valuables from safes without opening them and could enter a room without passing through any doors, windows, or chimneys. C. H. Hinton explained similar phenomena in "What Is the Fourth Dimension?" and other writings.

We can understand how such things can happen by making use of a fourth dimension that is intuitively accessible to us: time. However, this choice is made here only for purposes of illustration: Time is *a* fourth dimension, but it is not the only one that is conceivable mathematically or could be realized physically in a universe different from the one we happen to inhabit. "Objects" in our (3 + 1)-dimensional space-time are really *four*-dimensional; they possess duration as well as length, breadth, and height. A truly three-dimensional object would exist only for a single instant of time. Let us call that instant its *present*. The present forms a three-dimensional slice of space-time (Figure 30). It separates a four-dimensional region (the *past*) from another four-dimensional region (the *future*). All this is strictly analogous to a two-dimensional plane (Flatland) that separates a surrounding three-dimensional space into two three-dimensional regions (*above* and *below*).

The Sphere can steal A. Square's tablet of accounts from inside a locked cupboard because that

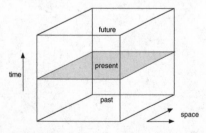

Figure 30 Past, present, and future in a four-dimensional space-time. Here space is drawn schematically as two-dimensional.

see also two tablets of accounts. I am about to descend into that cupboard and to bring you one of those tablets. I saw you lock the cupboard half an hour ago, and I know you have the key in your possession. But I descend from Space; the doors, you see, remain unmoved. Now I am in the cupboard and am taking the tablet. Now I have it. Now I ascend with it.

I rushed to the closet and dashed the door open. One of the tablets was gone.[2] With a mocking laugh, the Stranger appeared in the other corner of the room, and at the same time the tablet appeared upon the floor. I took it up. There could be no doubt — it was the missing tablet.

I groaned with horror, doubting whether I was not out of my senses; but the Stranger continued: "Surely you must now see that my explanation, and no other, suits the phenomena. What you call Solid things are really superficial; what you call Space is really nothing but a great Plane. I am in Space, and look down upon the insides of the things of which you only see the outsides. You could leave this Plane yourself, if you could but summon up the necessary volition. A slight upward or downward motion would enable you to see all that I can see.

"The higher I mount, and the further I go from your Plane, the more I can see, though of course I see it on a smaller scale. For example, I am ascending; now I can see your neighbour the Hexagon and his family in their several apartments;

now I see the inside of the Theatre, ten doors off, from which the audience is only just departing; and on the other side a Circle in his study, sitting at his books. Now I shall come back to you. And, as a crowning proof, what do you say to my giving you a touch, just the least touch, in your stomach? It will not seriously injure you, and the slight pain you may suffer cannot be compared with the mental benefit you will receive."

Before I could utter a word of remonstrance, I felt a shooting pain in my inside, and a demoniacal laugh seemed to issue from within me. A moment afterwards the sharp agony had ceased, leaving nothing but a dull ache behind, and the Stranger began to reappear, saying, as he gradually increased in size, "There, I have not hurt you much, have I? If you are not convinced now, I don't know what will convince you. What say you?"

My resolution was taken. It seemed intolerable that I should endure existence subject to the arbitrary visitations of a Magician who could thus play tricks with one's very stomach. If only I could in any way manage to pin him against the wall till help came!

Once more I dashed my hardest angle against him, at the same time alarming the whole household by my cries for aid. I believe, at the moment of my onset, the Stranger had sunk below our Plane, and really found difficulty in rising. In any case he remained motionless, while I, hearing, as I thought, the sound of some help approaching,

cupboard exists only in the plane and is impenetrable only from within that plane. From above or below, its boundary in the plane poses no obstacle. The Sphere merely displaces the tablet *upward* (and out of the plane), then moves it sideways (parallel to the plane), and finally takes it *downward* until it returns to the plane, but outside the cupboard (Figure 31a).

In the same way, a hypersphere could steal money from a locked three-dimensional safe. Being truly three-dimensional, the safe exists only in the present. The hypersphere displaces the safe's contents into the future, where the safe does not exist. Then it moves the money "parallel" to the present — that is, in space but not in time — until it is outside the place where the safe used to be. Finally, it moves the money back in the direction of the past, until it appears in the three-dimensional slice that is the present (Figure 31b).

If the safe also possesses duration in time, then a spatio-temporal hypersphere cannot use this technique. But if it has access to a fourth spatial dimension, it can use that dimension in the same way

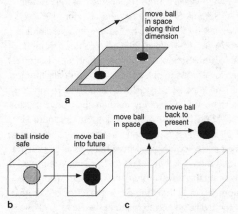

Figure 31 How extra-dimensional beings can steal objects from inside closed boxes. (**a**) The Sphere steals A. Square's tablets. (**b**) A hypersphere steals money from a locked three-dimensional safe. (**c**) If the safe has duration, a fourth spatial dimension is needed.

in which the above process uses time (Figure 31c). Rudy Rucker gives details in *The Fourth Dimension*.

3 One of the staples of science fiction and horror stories is the "creature from another dimension." The Sphere is one of the earliest characters in this tradition. He not only inhabits a dimension that is normally inaccessible to A. Square but also propels the unfortunate Flatlander out of his own universe into the Sphere's. This, too, is a common theme in the SF and horror literature; Abbott was one of the first to use it. C. H. Hinton took the speculative aspect much further, with numerous essays on such matters as ghosts being creatures from the fourth dimension and the existence of a four-dimensional God who was omniscient by virtue of seeing the entire three-dimensional world laid open before him. In his short story "An Unfinished Communication" (1895), the afterlife involves travel along the fourth dimension of time.

A key figure in this connection is H(erbert) G(eorge) Wells (1866–1946), who borrowed some of Hinton's ideas to explain how the central device in his celebrated novel *The Time Machine* (1895) worked — hence Wells's Time Traveller is a literary cousin of A. Square. SF author Stephen Baxter (1957–), an expert on Wells, has provided some fascinating details. In 1884 Wells, then age eighteen, attended the Normal School of Science, which later became the Royal College of Science and eventually merged with the Imperial College of Science and Technology. There he studied biology, mathematics, physics, geology, drawing, and astrophysics. He graduated in 1890 with "first class" honors in zoology and second class in geology. Somewhere during this period, he began the work that led up to *The Time Machine*. His first time-travel story, "The Chronic Argonauts," appeared in 1888 in the *Science Schools Journal*, which Wells helped to found at the Royal College. The protagonist voyages into the past and commits a murder. The story offers no rationale for time travel and is more of a mad-scientist tale in the tradition of Mary Shelley's *Frankenstein*. Wells later destroyed every copy of it that he could locate because it embarrassed him. It lacked even the paradoxical element

pressed against him with redoubled vigour, and continued to shout for assistance.

A convulsive shudder ran through the Sphere. "This must not be," I thought I heard him say: "either he must listen to reason, or I must have recourse to the last resource of civilization." Then, addressing me in a louder tone, he hurriedly exclaimed, "Listen: no stranger must witness what you have witnessed. Send your Wife back at once, before she enters the apartment. The Gospel of Three Dimensions must not be thus frustrated. Not thus must the fruits of one thousand years of waiting be thrown away. I hear her coming. Back! back! Away from me, or you must go with me — whither you know not — into the Land of Three Dimensions!"

"Fool! Madman! Irregular!" I exclaimed; "never will I release thee; thou shalt pay the penalty of thine impostures."

"Ha! Is it come to this?" thundered the Stranger: "then meet your fate: out of your Plane you go.[3] Once, twice, thrice! 'Tis done!"

of the 1891 "Tourmalin's Time Cheques" (also called "The Time Bargain") by F. Anstey (Thomas Anstey Guthrie, 1856–1934), which introduced many of the standard time-travel paradoxes. An example so well known that it has become a cliché is the Grandfather Paradox: If you travel back in time and kill your youthful grandfather, then you will not be born, and therefore you cannot travel back in time to kill him — but then you are born, so you can travel back....

Over the following three years, Wells produced two more versions of his time-travel story, now lost, but it seems that the story line mutated into a far-future vision of the human race. In 1894 he published the next version in the *National Observer* magazine, as three connected tales with the title *The Time Machine*. This version has many features in common with the final novel. Before the publication of Wells's series was complete, the editor of the magazine moved to another, *The New Review*. There he commissioned the same series again (it was published under the same title), but Wells made substantial changes. The manuscripts include many scenes that were never printed: The traveler journeys into the past, running into a prehistoric hippopotamus and meeting the Puritans in 1645. The published version is pretty much the one that appeared in book form in 1895. In this version, the Time Traveller moves only into the future, where he finds out what will happen to the human race, which speciates into the languid Eloi and the horrid Morlocks.

In a foreword to the 1932 edition, Wells says that he got the idea for the novel from "student discussions in the laboratories and debating society of the Royal College of Science in the eighties." According to Wells's son, the idea came from a paper on the fourth dimension read by another student. In the introduction to the story, the Time Traveller invokes the fourth dimension to explain why such a machine is possible:

"But wait a moment. Can an *instantaneous* cube exist?"

"Don't follow you," said Filby.

"Can a cube that does not last for any time at all, have a real existence?"

Filby became pensive.

"Clearly," the Time Traveller proceeded, "any real body must have extension in four directions: it must have Length, Breadth, Thickness, and — Duration....

"... There are really four dimensions, three which we call the three planes of Space, and a fourth, Time. There is, however, a tendency to draw an unreal distinction between the former three dimensions and the latter, because it happens that our consciousness moves intermittently in one direction along the latter from the beginning to the end of our lives....

"... But some philosophical people have been asking why *three* dimensions particularly — why not another direction at right angles to the three? — and have even tried to construct a Four-Dimensional geometry. Professor Simon Newcomb was expounding this to the New York Mathematical Society only a month or so ago."

Newcomb published on the topic of four-dimensional space from 1877, and he spoke about it to the New York Mathematical Society in 1893. Moreover, he was acquainted with C. H. Hinton; in fact, Newcomb probably secured Hinton his job at the Naval Observatory in Washington, D.C. in 1900. We don't know whether Wells met Hinton, but there is circumstantial evidence for some degree of influence. For example, the term scientific romance was coined by Hinton in titles of his collected speculative essays in 1884 and 1886, and Wells later used the same phrase to describe his own stories. Wells was a regular reader of *Nature*, which reviewed (favorably) Hinton's first series of "Scientific Romances" in 1885 and summarized several of his ideas on the fourth dimension.

Shortly afterward, *Nature* published a letter about the fourth dimension, signed enigmatically "S." It was a response to the review of Hinton's book, and it says in part

What is the fourth dimension?...I propose to consider time as the fourth dimension....Since this fourth dimension cannot be introduced into space ... we require a new kind of space for its existence, which we may call time-space.

Who was S? James E. Beichler suggests that it was the mathematician J.J. Sylvester, who along with his student William Kingdon Clifford (1845–1879) provided the original inspiration for Newcomb's own work. When Sylvester went to Johns Hopkins University, he met Newcomb, and the influence continued. Now in 1870, Sylvester had told his friend he and other mathematicians viewed time as a fourth dimension. Beichler states that Sylvester habitually signed letters to his friends as "S." However, Tim Axon of the H.G. Wells Society contacted Karen Parshall at the University of Virginia, as expert on Sylvester's correspondence. She told him that Sylvester did not habitually sign his name as "S." Indeed no examples exist, and when he signed something other than J.J. Sylvester, it was always "J.J.S." Moreover, Sylvester's surviving correspondence from 1885 has no mention of the *Nature* letter. However, Sylvester did occasionally write about the fourth dimension.

If Sylvester did write the *Nature* letter, we can credit him with inspiring both *Flatland* and *The Time Machine*, probably via Hinton in both cases. But Parshall felt that the letter didn't follow Sylvester's usual style, so it may well have been written by someone else. One possibility is W. Irving Stringham, one of Sylvester's best students at Johns Hopkins University. Stringham specialized in four-dimensional geometry, publishing three-dimensional cross sections of four-dimensional polytopes in the *American Journal of Mathematics* in 1880. After graduation, Stringham went to Leipzig to work with Felix Klein, one of the world leaders in multidimensional geometry. They persuaded Victor Schlegel (another "S" but much less plausible as the author of the *Nature* letter) to make a series of three-dimensional models related to four-dimensional polytopes. The models were displayed in 1884—the year before the *Nature* letter was published—and were subsequently sold commercially. So the timing is consistent, but the evidence for Stringham as "S" is circumstantial.

SF writers since Wells have also made effective use of a fourth dimension. Modern time-travel books include *The End of Eternity* (1955) by Isaac Asimov and *Up the Line* (1969) by Robert Silverberg (1935–), both of which explore the paradoxes of changing the past; *The Man Who Folded Himself* (1973) by David Gerrold (Jerrold David Friedman, 1944–), which takes the paradoxes to serious extremes; and *Pastwatch: The Redemption of Christopher Columbus* (1996) by Orson Scott Card (1951–), in which people from the future rewrite the history of America. Of course, a fourth dimension need not represent time. In the 1949 *Skylark of Valeron* (originally serialized in magazine form in 1934) by E(dward) E(lmer) Smith (1890–1965), the protagonists make a short excursion into a fourth dimension. Smith's celebrated "Lensman" series, a six-volume epic about a cosmic battle between the good Arisians and the evil Eddorians, makes use of several extra-dimensional concepts. All nonaqueous, cold-blooded non-oxygen breathers, such as the Palainians, the evil Eich, and the Onlonians, have a metabolism that extends into the fourth dimension (*First Lensman,* 1950; *Second Stage Lensmen,* 1941–2/1954; *Children of the Lens,* 1947–8/1954). (Here, if two dates are given, the first is the magazine serialization and the second the first book publication.) The hyperspatial tube, invented by the Eddorians (*Triplanetary,* 1934/1948; *Galactic Patrol,* 1937/1950), allows the use of a fourth dimension to bypass obstacles in ordinary space. It has evident military value and is used by the Eich and the Overlords of Delgon (*Gray Lensman,* 1939–40/1951) to bring terror to the space lanes. It also plays a major role in several subsequent battles for the fate of civilization.

In the horror genre, the "Cthulhu" works of H(oward) P(hillips) Lovecraft (1890–1937) make extensive use of extra-dimensional beings, although most of these are demons or monsters, who every so often materialize out of the walls or in mid-air and drag people off into a hitherto unsuspected dimension. Pratchett makes similar use of the "dungeon dimensions"—an infinite wasteland outside space and time.

§ 18. — How I came to Spaceland, and what I saw there.[1]

An unspeakable horror seized me. There was a darkness; then a dizzy, sickening sensation of sight that was not like seeing; I saw a Line that was no Line;[2] Space that was not Space: I was myself, and not myself. When I could find voice, I shrieked aloud in agony, "Either this is madness or it is Hell." "It is neither," calmly replied the voice of the Sphere, "it is Knowledge; it is Three Dimensions: open your eye once again and try to look steadily."

I looked, and, behold, a new world![3] There stood before me, visibly incorporate, all that I had before inferred, conjectured, dreamed, of perfect Circular beauty. What seemed the centre of the Stranger's form lay open to my view: yet I could see no heart, nor lungs, nor arteries, only a beautiful harmonious Something — for which I had no words; but you, my Readers in Spaceland, would call it the surface of the Sphere.

Prostrating myself mentally before my Guide,[4] I cried, "How is it, O divine ideal of consummate loveliness and wisdom that I see thy inside, and yet cannot discern thy heart, thy lungs, thy arteries,

1 The discussion of what spatial dimensions an evolved brain will be able to perceive makes it highly unlikely that A. Square, whose brain evolved in Flatland, would actually be able, when transported into Spaceland, to perceive solid objects *the same way we do.* But it would spoil a good story to worry about such details. (The same narrative difficulty arises in *Flatterland,* where it is "solved" by inventing the VUE — Virtual Unreality Engine — which interfaces directly with the heroine's brain and gives her direct perceptual access to all possible mathematical spaces. It is highly unlikely that a real VUE for humans could ever be constructed — our brains lack the wiring for it — but so what?)

2 In a single sentence Abbott manages to convey A. Square's sense of disorientation in his new surroundings. Similar techniques were later employed by other authors, especially David Zindell in *Neverness* (1988) and its sequels:

> If this tremendously concentrated thinking caused the manifold to distort into a series of infinite trees or to warp into a Danladi bubble — well, that was interesting, but not nearly so interesting as the openness or closure of realspace.... She [the Solid State Entity] knew that certain pockets of the manifold were distorted in certain ways. This knowledge saved some of us pilots from stumbling into infinite trees, as I once had. She warned us away from the worst dangers. She provided us with mappings ... and she provided us with the fixed-points of Gehenna Luz [a star on the verge of exploding].

The hero Mallory Ringess, voyaging inside the Solid State Entity, determines to prove the Great Theorem that will unlock the deepest secrets of interstellar travel:

> And then the correspondence scheme collapsed, almost as the layers of a star collapse around the core when it becomes a supernova, and there was a choosable mapping. *There was a choosable mapping!* There was elegance, beauty, and starlight.

I made a mapping. The white light of dreamtime swept by in brilliant streamers, then collapsed into a single point of starlight which burned and expanded and brightened until it filled all of my mind.... I fell out in realspace above the hot white star that the Entity had named Gehenna Luz. I had proved the Great Theorem, I had journeyed far at a single fall, and now all the stars in the sky were finally mine.

3 The "new" world here contains the old one, Flatland, and its novelty is its extra dimension. Today's mathematicians take this metaphor literally, and in applied mathematics, the appearance of an innovation is often consciously modeled as movement into a new dimension. In theoretical ecology, for instance, the populations of various creatures are considered as coordinates in an ecological "phase space." For example, the numbers of rabbits and foxes in a wood can be considered as coordinates in a plane. If there are r rabbits and f foxes, then the state of the wood is represented by the point with coordinates (r, f). Ecologists write down mathematical rules for how r and f change over time, and in the absence of any external influences, the point (r, f), which represents the state of the ecosystem, always remains within the same plane, just as A. Square normally remains inside *his* plane. The introduction of a new species — say, pheasants, whose population is represented by a new coordinate, p — changes the rules of the ecosystem, because foxes can now prey on pheasants as well as on rabbits. The new state is represented by a point (r, f, p) in *three*-dimensional space. A new dimension has appeared: the dimension "pheasant population." Just as the Sphere propels A. Square into a new dimension, so the introduction of pheasants propels the woodland ecology into a new dimension. Similar mathematical models are used to study (in economics) the introduction of a new product into a market and (in biology) the evolution of a new species.

thy liver?" "What you think you see, you see not," he replied; "it is not given to you, nor to any other Being, to behold my internal parts. I am of a different order of Beings from those in Flatland. Were I a Circle, you could discern my intestines, but I am a Being, composed as I told you before, of many Circles, the Many in the One, called in this country a Sphere. And, just as the outside of a Cube is a Square, so the outside of a Sphere presents the appearance of a Circle."

Bewildered though I was by my Teacher's enigmatic utterance, I no longer chafed against it, but worshipped him in silent adoration. He continued, with more mildness in his voice. "Distress not yourself if you cannot at first understand the deeper mysteries of Spaceland. By degrees they will dawn upon you. Let us begin by casting back a glance at the region whence you came. Return with me a while to the plains of Flatland, and I will shew you that which you have often reasoned and thought about,[5] but never seen with the sense of sight—a visible angle." "Impossible!" I cried; but, the Sphere leading the way, I followed as if in a dream, till once more his voice arrested me: "Look yonder, and behold your own Pentagonal house, and all its inmates."

I looked below, and saw with my physical eye all that domestic individuality which I had hitherto merely inferred with the understanding. And how poor and shadowy was the inferred conjecture in comparison with the reality which I now

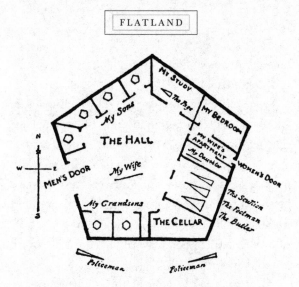

beheld! My four Sons calmly asleep in the North-Western rooms, my two orphan Grandsons to the South; the Servants, the Butler, my Daughter, all in their several apartments. Only my affectionate Wife, alarmed by my continued absence, had quitted her room and was roving up and down in the Hall, anxiously awaiting my return. Also the Page, aroused by my cries, had left his room, and under pretext of ascertaining whether I had fallen somewhere in a faint, was prying into the cabinet in my study. All this I could now *see*, not merely infer; and as we came nearer and nearer, I could discern even the contents of my cabinet, and the two chests of gold, and the tablets of which the Sphere had made mention.

Touched by my Wife's distress, I would have sprung downward to reassure her, but I found myself incapable of motion. "Trouble not yourself about your Wife," said my Guide: "she will not be

4 In an unpublished manuscript titled *Confessions of a Rationalist Christian*, Abbott says that "I read with my older pupils Shakespeare, *In Memoriam*, Milton, Spenser, Bacon, once also a translation of Dante, and Pope regularly." The most celebrated work of Dante Alighieri (1265–1321) is *La Divina Commedia* (*The Divine Comedy*, originally just *La Commedia*), which was probably written between 1308 and 1321. It is an allegorical/polemical work in three parts: *Inferno* (*Hell*), *Purgatorio* (*Purgatory*), and *Paradiso* (*Paradise*). Dante appears as his own protagonist and is guided through Hell and Purgatory by the ghost of the Roman poet Virgil (70 B.C.–19 B.C.) and through Heaven by the fictional beloved, Beatrice. The Sphere plays the same role of guide and mentor for A. Square, for similar narrative reasons: Someone or something has to know what is happening and explain it to the protagonist so that the reader can follow the action.

5 A. Square's ability to perceive two-dimensional images of three-dimensional objects is not very plausible because his visual sense is structured to perceive one-dimensional images of two-dimensional objects. Endowing him with this ability, however, allows Abbott to get on with explaining to his readers the effects of projecting three-dimensional objects onto a two-dimensional surface, as the human eye does when it creates an image on the retina and as a camera does when it creates an image on film.

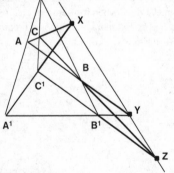

Figure 32 Perspective makes distant objects look smaller.

6 Thanks to perspective, the projection of an object onto a plane becomes smaller as the object becomes more distant; see Figure 32. This is why we can cover the moon with a thumb held at arm's length, for example. The geometry of such projections (reasonably known as *projective geometry*) is a beautiful branch of classical mathematics — it dates mainly from the nineteenth century. It refers to lines and points, but not to angles or distances. A typical theorem is Desargues's theorem (named after Girard Desargues, 1591–1661): If one triangle is a projection of another, then the intersections of corresponding pairs of sides all lie on a straight line (Figure 33). One interesting (though not especially important) application of projective geometry makes it possible to work out where a camera was placed from the photograph it took, by measuring the positions of just five features of the photograph and comparing them with reality. See Ian Stewart, *Visions Géométriques.*

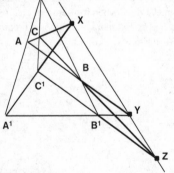

Figure 33 Desargues' theorem: If one triangle A'B'C' is a projection of a second triangle ABC, then the intersections X, Y, Z, of corresponding sides lie on a straight line.

long left in anxiety; meantime, let us take a survey of Flatland."

Once more I felt myself rising through space. It was even as the Sphere had said. The further we receded from the object we beheld, the larger became the field of vision. My native city, with the interior of every house and every creature therein, lay open to my view in miniature.[6] We mounted higher, and lo, the secrets of the earth, the depths of mines and inmost caverns of the hills, were bared before me.

Awestruck at the sight of the mysteries of the earth, thus unveiled before my unworthy eye, I said to my Companion, "Behold, I am become as a God. For the wise men in our country say that to see all things, or as they express it, *omnividence,* is the attribute of God alone." There was something of scorn in the voice of my Teacher as he made answer: "Is it so indeed? Then the very pick-pockets and cut-throats of my country are to be worshipped by your wise men as being Gods: for there is not one of them that does not see as much as you see now. But trust me, your wise men are wrong."

I. Then is omnividence the attribute of others besides Gods?

Sphere. I do not know. But, if a pick-pocket or a cut-throat of our country can see everything that is in your country, surely that is no reason why the pick-pocket or cut-throat should be accepted by you as a God. This omnividence, as you call it—

it is not a common word in Spaceland—does it make you more just, more merciful, less selfish, more loving? Not in the least. Then how does it make you more divine?

I. "More merciful, more loving!" But these are the qualities of women! And we know that a Circle is a higher Being than a Straight Line, in so far as knowledge and wisdom are more to be esteemed than mere affection.

Sphere. It is not for me to classify human faculties according to merit. Yet many of the best and wisest in Spaceland think more of the affections than of the understanding,[7] more of your despised Straight Lines than of your belauded Circles. But enough of this. Look yonder. Do you know that building?

I looked, and afar off I saw an immense Polygonal structure, in which I recognized the General Assembly Hall of the States of Flatland, surrounded by dense lines of Pentagonal buildings at right angles to each other, which I knew to be streets; and I perceived that I was approaching the great Metropolis.

"Here we descend," said my Guide. It was now morning, the first hour of the first day of the two thousandth year[8] of our era. Acting, as was their wont, in strict accordance with precedent, the highest Circles of the realm were meeting in solemn conclave, as they had met on the first hour of the first day of the year 1000, and also on the first hour of the first day of the year 0.[9]

7 Even though Abbott was a strong advocate of female education, here he slips back into a more conventional Victorian way of thinking that has by no means disappeared, even today. In effect, he is saying that Flatland should value its Women for their emotional attributes, not for their intelligence. But perhaps, given the poor state of female education in his day, he was merely trying to point out that intelligence is not the only sensible criterion for judging the worth of a human being. *If so,* he was pointing out the value of "emotional intelligence" a full century before Daniel Coleman.

8 In Flatland, the start of a new millennium is of high significance for the Priesthood, hence for the entire population, and must be marked by a solemn ceremony. It is much the same in Spaceland, though few of the recent millennial celebrations were solemn.

9 Abbott gives few details of the Flatland calendar, but it is interesting to see this confirmation that in at least one respect, it is more logical than ours.

10 Why Abbott chooses to emphasize the *perfect* symmetry of this particular square here is unclear. Mathematically, all squares are symmetric under the appropriate eight symmetries, or they are not true squares. Is A. Square's brother *more* symmetric than A. Square himself (making him worthy of such an important job as Chief Clerk of the High Council)? If so, the alleged regularity of Flatland's Polygonal classes may be less exact than A. Square has led us to believe. The context does seem to support the view that A. Square's brother is a little better than most ordinary Squares.

The minutes of the previous meetings were now read by one whom I at once recognized as my brother, a perfectly Symmetrical Square,[10] and the Chief Clerk of the High Council. It was found recorded on each occasion that: "Whereas the States had been troubled by divers ill-intentioned persons pretending to have received revelations from another World, and professing to produce demonstrations whereby they had instigated to frenzy both themselves and others, it had been for this cause unanimously resolved by the Grand Council that on the first day of each millenary, special injunctions be sent to the Prefects in the several districts of Flatland, to make strict search for such misguided persons, and without formality of mathematical examination, to destroy all such as were Isosceles of any degree, to scourge and imprison any regular Triangle, to cause any Square or Pentagon to be sent to the district Asylum, and to arrest any one of higher rank, sending him straightway to the Capital to be examined and judged by the Council."

"You hear your fate," said the Sphere to me, while the Council was passing for the third time the formal resolution. "Death or imprisonment awaits the Apostle of the Gospel of Three Dimensions." "Not so," replied I, "the matter is now so clear to me, the nature of real space so palpable, that methinks I could make a child understand it. Permit me but to descend at this moment and enlighten them." "Not yet," said my Guide, "the time will come for that. Meantime I must perform my mission. Stay thou there in thy place." Saying these words, he leaped with great dexterity into the sea (if I may so call it) of Flatland, right in the midst of the ring of Counsellors. "I come," cried he, "to proclaim that there is a land of Three Dimensions."

I could see many of the younger Counsellors start back in manifest horror, as the Sphere's circular section widened before them. But on a sign from the presiding Circle—who shewed not the slightest alarm or surprise—six Isosceles of a low type from six different quarters rushed upon the Sphere. "We have him," they cried; "No; yes; we have him still! he's going! he's gone!"

"My Lords," said the President to the Junior Circles of the Council, "there is not the slightest need for surprise; the secret archives, to which I alone have access, tell me that a similar occurrence happened on the last two millennial commencements.[11] You will, of course, say nothing of these trifles outside the Cabinet."

Raising his voice, he now summoned the guards. "Arrest the policemen; gag them. You know your duty." After he had consigned to their fate the wretched policemen—ill-fated and unwilling witnesses of a State-secret which they were not to be permitted to reveal[12]—he again addressed the Counsellors. "My Lords, the business of the Council being concluded, I have only to wish you a happy New Year." Before departing,

11 Further evidence that new millennia in Flatland —unlike those in Spaceland, but in accordance with the erroneous assumptions of Spaceland popular culture—begin in years ending with 00. According to the *Oxford English Dictionary* (*OED*), a century is (italics as in the original)

> Each of the successive periods of 100 years, reckoning from a received chronological epoch; *especially* from the *assumed* date of the birth of Christ: thus the hundred years from that date to the year A.D. 100 were the *first century* of the Christian Era; those from 1801 to 1900 inclusive are the *nineteenth century.*

Century, of course, has other meanings, but this is the relevant one. (The *OED* is less specific about *millennium,* in part because to many Christians, *The Millennium* refers to Christ's prophesied 1000-year reign, after his return to Earth in the Second Coming.) Among the *OED*'s illustrative quotations is one from Charles Knight's *Passages of a Working Life During Half a Century,* published 1864–1865: "The learned had settled, after a vast deal of popular controversy, that the century had its beginning on the 1st of January, 1801, and not on the 1st of January, 1800."

Clearly, the issue caused as much heated discussion in the countdown to 1800 as it did in that to 2000. A sensible way to resolve the numerological dilemma (as 2000 approached, several national governments recognized this possibility but chickened out) is to redefine the calendar so that new centuries begin in years ending 00, not 01. In Flatland this more logical system prevails. The new millennium starts in 2000 (p. 131, the end of 1999 is "the last time in the old Millennium"; p. 133, "The third Millennium had begun"). Moreover, there is explicit reference to the year 0. Abbott was writing in the early 1880s, with the new century already looming on the horizon; perhaps he was quietly advocating a more logical solution to the calendrical conundrum. However, he did not go as far as he might have. If centuries begin with 00 and end with 99, then with conventional terminology the first century should start in the year 0 and end in A.D. 99, and the twentieth century should start in 1900 and end in 1999. However, if we started counting *years* from 0, it would make sense to start counting *centuries* from 0 too, so that the period from 0 to A.D. 99 would be the zeroth century, and that from 1900 to 1999 would be the nineteenth (and would coincide with "the 1900s"). By the same token, the period from 0 to A.D. 999 is the zeroth millennium, and that from 2000 to 2999 is the second millennium, not the third. Anyone who advocates 2000 as the start of the new millennium should, by the same logic, consider it the start of the twentieth century, not the twenty-first.

Oddly, the standard usage for decades is not in dispute, even though it is inconsistent with that for centuries: In the twentieth century, *the sixties* means 1960–1969 and *the nineties* means 1990–1999. When, in 1999, readers of *The Guardian* newspaper in the UK were challenged to find a name for 2000–2009, the cleverest proposal was "the Noughties." This resonates with both the zeroth decade (*nought* means "zero") and with the "Naughty Nineties," the period of 1890–1899.

12 Abbott was probably alluding to the power of the civil servants (government employees) who ran the British Empire: The State had many secrets, kept firmly under wraps to make sure that ordinary Britons had as little idea as possible of the things being done by the ruling classes. Today, Britain still has one of the most restrictive attitudes toward official secrets in the Western world, and nothing that resembles American laws ensuring freedom of information exists there. Secrecy begets paranoia, and conspiracy theories abound even in the United States, with allegations of official cover-ups of everything from the assassination of John F. Kennedy to alien invasions. Indeed, many modern Americans (and others) believe that the United States government has for many years been in contact with aliens.

he expressed, at some length, to the Clerk, my excellent but most unfortunate brother, his sincere regret that, in accordance with precedent and for the sake of secrecy, he must condemn him to perpetual imprisonment, but added his satisfaction that, unless some mention were made by him of that day's incident, his life would be spared.

This conspiracy theory dates principally from the Roswell incident, in which parts of a strange "craft" landed in a field near Roswell, New Mexico, sometime during the first week of July 1947. Roswell is near a huge U.S. Air Force base. A New Mexican rancher, W. W. ("Mac") Brazel, rode out with his neighbors' son and discovered some strange debris. A few days later he reported his find to Sheriff George Wilcox, who passed the information on to Major Jesse Marcel of the 509 Bomb Group. On July 8, the commander of the 509 Bomb Group issued a press release to the effect that wreckage of a disk-shaped vehicle/aircraft that had crashed to the ground had been recovered. A few hours later a new press release contradicted the first, saying that a weather balloon had mistakenly been identified as wreckage of a flying saucer. The Roswell incident played a major role in starting the UFO (unidentified flying object) craze, and many views about it can easily be found on the Internet. It gave rise to the theory that the U.S. government was secretly in contact with aliens and that alien corpses had been recovered from the wreck and taken to the U.S. Air Force's secret Area 51 installation at Groom Lake, Nevada — where, for example, the U-2 spy plane and the F-117 stealth fighter were tested during development. Later, film purporting to show the dissection of the aliens became public.

Dave Thomas, writing in the January-February 1995 issue of the *Skeptical Inquirer,* gives strong evidence for a more prosaic theory. There was indeed a government cover-up, and the weather balloon story was untrue. The wreckage was actually that of a balloon flight forming part of Project Mogul, a top-secret project to monitor Russian nuclear weapons using high-altitude microphones. One such train of balloons, flight number 4, was launched on 4 July 1947, and components of this balloon train account for much of what seems to have been found on the ground. Moreover, evidence has turned up to support the contention that the "alien dissection" movie was a deliberate fake.

To UFO aficionados, the Roswell incident is the best-documented example of an alien presence on Earth (UFO disbelievers concur), and they reject all official explanations. It is certainly possible that the Project Mogul explanation is not the whole truth. For example, the project might have been an *illegal* experiment involving, say, the scattering of disease organisms or radioactive material. Other illegal experiments by branches of the U.S. and UK governments have surfaced in recent years. But aliens it wasn't.

In Flatland, however, the cover-up is genuine, and the alien is real.

§ 19. — How, though the Sphere shewed me other mysteries of Spaceland, I still desired more; and what came of it.

WHEN I saw my poor brother led away to imprisonment, I attempted to leap down into the Council Chamber, desiring to intercede on his behalf, or at least bid him farewell. But I found that I had no motion of my own. I absolutely depended on the volition of my Guide, who said in gloomy tones, "Heed not thy brother; haply thou shalt have ample time hereafter to condole with him. Follow me."

Once more we ascended into space. "Hitherto," said the Sphere, "I have shewn you naught save Plane Figures and their interiors. Now I must introduce you to Solids, and reveal to you the plan upon which they are constructed. Behold this multitude of moveable square cards. See, I put one on another, not, as you supposed, Northward of the other, but *on* the other. Now a second, now a third. See, I am building up a Solid by a multitude of Squares parallel to one

(1)

(2)

another. Now the Solid is complete, being as high as it is long and broad, and we call it a Cube."

"Pardon me, my Lord," replied I; "but to my eye the appearance is as of an Irregular Figure[1] whose inside is laid open to the view; in other words, methinks I see no Solid, but a Plane such as we infer in Flatland; only of an Irregularity which betokens some monstrous criminal, so that the very sight of it is painful to my eyes."

"True," said the Sphere; "it appears to you a Plane, because you are not accustomed to light and shade and perspective; just as in Flatland a Hexagon would appear a Straight Line to one who has not the Art of Sight Recognition. But in reality it is a Solid, as you shall learn by the sense of Feeling."

He then introduced me to the Cube, and I found that this marvellous Being was indeed no Plane, but a Solid; and that he was endowed with six plane sides and eight terminal points called solid angles; and I remembered the saying of the Sphere that just such a Creature as this would be formed by a Square moving, in Space, parallel to himself: and I rejoiced to think that so insignificant a Creature as I could in some sense be called the Progenitor of so illustrious an offspring.

But still I could not fully understand the meaning of what my Teacher had told me concerning "light" and "shade" and "perspective"; and I did not hesitate to put my difficulties before him.

Were I to give the Sphere's explanation of these

1 A. Square is perceiving the *projection* into two dimensions of a three-dimensional cube, just as we do, but — no doubt for purposes of instruction — Abbott makes his protagonist "[un]accustomed to light and shade and perspective," some of the cues that the human brain uses to deduce three-dimensional relationships from two-dimensional images. Our intuition for three dimensions is actually very poor (one of the world's leading experts in three-dimensional topology once stated that he had no intuition for three dimensions at all). In chapter 3 of *Flatterland*, "The Visitation," I mention a vivid illustration of this: A hole can be made in a cube that permits a *larger* cube (6% larger, in fact) to pass through. This puzzle is known as Prince Rupert's Cubes (Figure 34). Here is another illustration: Try to visualize the sequence of cross sections formed when a cube passes through a plane in the direction of its main diagonal. (Put the book down *now,* and give it a try.)

In fact, the sequence of cross sections begins with a single point, which turns into an equilateral triangle. The triangle grows, and when it reaches a critical size, it begins to lose small triangles at its corners. The missing portions become larger relative to the triangle, until (exactly half-way through) the cross section becomes a regular hexagon. Now the whole sequence reverses, but rotated by 180°. The sequence (Figure 35) is highly unintuitive to most people: Abbott was wise to choose a sphere

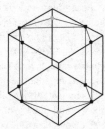

Figure 34 Each black dot is a quarter of the way along from the nearest corner; each gray dot is $3/16$ of the way along from the nearest corner. The hole that they define is big enough to accommodate passage of a cube 1.06 times as big as the original cube.

Figure 35 Slices of a cube perpendicular to its long diagonal.

as his visitor from the third dimension, rather than a tilted cube. As we have noted, it is unlikely that the visual system of a two-dimensional being would actually behave in the way Abbott's narrative requires. On the other hand, A. Square's sense of Feeling, which Abbott describes as being used to convince him of the true form of a cube, would probably work well, although he would "feel" various planar sections of the cube rather than the whole thing at once.

2 Prometheus appears, along with his brother Epimetheus, in the Olympian creation myths of the ancient Greeks. The names mean, respectively, "forethought" and "afterthought." Prometheus is often credited with the creation of humankind, and he stole fire from Olympus and gave it to humans (the act to which Abbott is alluding). Prometheus had supported Zeus in his war with the Titans, but his theft of fire angered Zeus, who punished him by chaining him to a crag in the Caucasus mountains. Here an eagle tore all day at his liver, which grew back every night. After many generations, Heracles shot the eagle — with Zeus's agreement — and freed the rebellious hero. Aeschylus's tragedy *Prometheus Bound* is about the hero's dreadful punishment; its lost sequel *Prometheus Unbound* told of his rescue. Percy Bysshe Shelley (1792–1822) wrote a dramatic poem with the same title as the sequel, in which Prometheus represents those who challenge tyranny for humanity's benefit. Abbott, as a lover of poetry, had probably read Shelley, but he would have known about Prometheus in any case, from his education in the classics.

3 But why stop with a finite number? There is no logical reason to "limit our Dimensions" to finite val-

matters, succinct and clear though it was, it would be tedious to an inhabitant of Space, who knows these things already. Suffice it, that by his lucid statements, and by changing the position of objects and lights, and by allowing me to feel the several objects and even his own sacred Person, he at last made all things clear to me, so that I could now readily distinguish between a Circle and a Sphere, a Plane Figure and a Solid.

This was the Climax, the Paradise, of my strange eventful History. Henceforth I have to relate the story of my miserable Fall:—most miserable, yet surely most undeserved! For why should the thirst for knowledge be aroused, only to be disappointed and punished? My volition shrinks from the painful task of recalling my humiliation; yet, like a second Prometheus,[2] I will endure this and worse, if by any means I may arouse in the interiors of Plane and Solid Humanity a spirit of rebellion against the Conceit which would limit our Dimensions to Two or Three or any number short of Infinity.[3] Away then with all personal considerations! Let me continue to the end, as I began, without further digressions or anticipations, pursuing the plain path of dispassionate History. The exact facts, the exact words,—and they are burnt in upon my brain,—shall be set down without alteration of an iota; and let my Readers judge between me and Destiny.

The Sphere would willingly have continued his lessons by indoctrinating me in the conformation of all regular Solids,[4] Cylinders, Cones, Pyramids, Pentahedrons, Hexahedrons, Dodecahedrons, and Spheres: but I ventured to interrupt him. Not that I was wearied of knowledge. On the contrary, I thirsted for yet deeper and fuller draughts than he was offering to me.

"Pardon me," said I, "O Thou Whom I must no longer address as the Perfection of all Beauty; but let me beg thee to vouchsafe thy servant a sight of thine interior."

Sphere. My what?

I. Thine interior: thy stomach, thy intestines.

Sphere. Whence this ill-timed impertinent request? And what mean you by saying that I am no longer the Perfection of all Beauty?

I. My Lord, your own wisdom has taught me to aspire to One even more great, more beautiful, and more closely approximate to Perfection than yourself. As you yourself, superior to all Flatland forms, combine many Circles in One, so doubtless there is One above you who combines many Spheres in One Supreme Existence, surpassing even the Solids of Spaceland. And even as we, who are now in Space, look down on Flatland and see the insides of all things, so of a certainty there is yet above us some higher, purer region, whither thou dost surely purpose to lead me — O Thou Whom I shall always call, everywhere and in all Dimensions, my Priest, Philosopher, and Friend — some yet more spacious Space, some more dimensionable Dimensionality, from the vantage-

ues, and today's mathematics acknowledges this by including a vast array of infinite-dimensional spaces. Perhaps the best-known example is Hilbert Space, the mathematical substrate upon which quantum theory rests, so as far as mathematical physics goes, everything in the universe is a pattern in an infinite-dimensional space. In its simplest guise, Hilbert Space consists of all infinite sequences

$$x_0, x_1, x_2, x_3, \ldots$$

for which the infinite series

$$x_0^2 + x_1^2 + x^2 + x_3^2 + \ldots$$

converges (has a finite sum). Observe the close analogy with n-dimensional Euclidean space. Each coordinate x_i determines a distinct "direction" in this space, and there are infinitely many coordinates and hence infinitely many dimensions. A more familiar example of infinite-dimensional spaces is the space of all functions $f(x)$ of a real variable x. An entire subject, functional analysis, has grown out of the need to have a working theory of infinite-dimensional spaces.

In *The Fourth Dimension,* Rudy Rucker remarks that the hyperspace philosophers of the late nineteenth century were aware that because two-dimensional space is a part of three-dimensional space, three-dimensional space is a part of four-dimensional space, and so on, there is an infinite regress. "Where can it stop? Only at infinity." C. H. Hinton compared this infinite regress to the theory that the world stands on the back of a tortoise. On what does the tortoise stand? Another tortoise. And....The opening paragraph of the bestseller *A Brief History of Time* by Stephen Hawking (1942–) relates the tale of a public lecturer, possibly Bertrand Russell (1872–1970), who had been talking on astronomy:

> At the end of the lecture, a little old lady at the back of the room got up and said: "What you have told us is rubbish. The world is really a flat plate supported on the back of a giant tortoise." The scientist gave a superior smile before replying, "What is the tortoise standing on?" "You're very clever, young man, very clever," said the old lady. "But it's turtles all the way down!"

In *Figments of Reality,* Jack Cohen and I point out the following curious parallel in the problem of "what happened before the Big Bang":

> What is rather odd about this particular question (the origin of time) is that most people seem perfectly happy with "it's always been like that," finding no difficulty in conceiving of a universe that goes back forever.
>
> Why shouldn't they?
>
> Well....We opened this chapter with the story of the old lady who argued that the only way to support the Earth would be "turtles all the way down." Nearly everybody finds an infinite pile of turtles highly incongruous — as an explanation it's a non-starter. So why are we so happy with an infinite pile of causality: today's universe riding on the back of yesterday's, which rides in turn on the day before's?
>
> *It's universes all the way back!*
>
> Mathematicians, whose understanding of the slippery nature of the infinite is more fully developed, feel that a universe that has always existed needs just as much explaining as one that suddenly pops into being from nowhere (and no-time). An entirely reasonable mathematician's question about a system that goes back forever is: "Yes, but where did it *all* come from?" Infinity is not an impassable barrier, it's just a name that sums up a way of thinking about recursive processes.

In 1893 Arthur Willink, a Victorian theologian and one of the hyperspace philosophers, published *The World of the Unseen,* in which he suggested that infinite-dimensional space was exactly the kind of place where God would live:

> When we have recognized the existence of Space of Four Dimensions there is no greater strain called for in the recognition of the existence of Space of Five Dimensions, and so on up to a Space of an infinite number of Dimensions.... [T]o an eye in the Highest Space of all, an infinitely perfect revealing of the most hidden and secret things is necessarily presented. This emphasizes very strongly what has been said

ground of which we shall look down together upon the revealed insides of Solid things, and where thine own intestines, and those of thy kindred Spheres, will lie exposed to the view of the poor wandering exile from Flatland, to whom so much has already been vouchsafed.

Sphere. Pooh! Stuff! Enough of this trifling! The time is short, and much remains to be done before you are fit to proclaim the Gospel of Three Dimensions to your blind benighted countrymen in Flatland.

I. Nay, gracious Teacher, deny me not what I know it is in thy power to perform. Grant me but one glimpse of thine interior, and I am satisfied for ever, remaining henceforth thy docile pupil, thy unemancipable slave, ready to receive all thy teachings and to feed upon the words that fall from thy lips.

Sphere. Well, then, to content and silence you, let me say at once, I would shew you what you wish if I could; but I cannot. Would you have me turn my stomach inside out to oblige you?

I. But my Lord has shewn me the intestines of all my countrymen in the Land of Two Dimensions by taking me with him into the Land of Three. What therefore more easy than now to take his servant on a second journey into the blessed region of the Fourth Dimension,[5] where I shall look down with him once more upon this land of Three Dimensions, and see the inside of every three-dimensioned house, the secrets of the solid earth, the treasures of the mines in Spaceland, and the intestines of every solid living creature, even of the noble and adorable Spheres.

Sphere. But where is this land of Four Dimensions?

I. I know not: but doubtless my Teacher knows.

Sphere. Not I. There is no such land. The very idea of it is utterly inconceivable.

I. Not inconceivable, my Lord,[6] to me, and therefore still less inconceivable to my Master. Nay, I despair not that, even here, in this region of Three Dimensions, your Lordship's art may make the Fourth Dimension visible to me; just as in the Land of Two Dimensions my Teacher's skill would fain have opened the eyes of his blind servant to the invisible presence of a Third Dimension, though I saw it not.

Let me recall the past. Was I not taught below that when I saw a Line and inferred a Plane, I in reality saw a Third unrecognized Dimension, not the same as brightness, called "height"? And does it not now follow that, in this region, when I see a Plane and infer a Solid, I really see a Fourth unrecognized Dimension,[7] not the same as colour, but existent, though infinitesimal and incapable of measurement?

And besides this, there is the Argument from Analogy of Figures.

Sphere. Analogy! Nonsense: what analogy?

I. Your Lordship tempts his servant to see whether he remembers the revelations imparted to him. Trifle not with me, my Lord; I crave, I

about the Omniscience of God. For He, dwelling in the Highest Space of all, not only has this perfect view of all the constituents of our being, but also is most infinitely near to every point and particle of our whole constitution.

If only it were that simple.... Thanks to the highly original work of the German mathematician Georg Cantor (1845–1918), we now know that mathematics possesses an endless hierarchy of infinities, each bigger than the previous one. Cantor's starting point is the idea that two finite sets have the same number of members if and only if they can be put into one-to-one correspondence. If a table is set so that every place has one knife and one fork, then we know, without counting them, that in total there are as many knives as forks. Cantor extended this property to infinite sets, making it the *definition* of "transfinite cardinals," the infinite analogue of finite numbers. The most obvious transfinite cardinal is that of the integers 1, 2, 3, 4, …, which he symbolized as $\aleph_0$ (aleph-null, aleph being the first letter of the Hebrew alphabet). He then proved that the set of all real numbers (infinite decimals) cannot be put into one-to-one correspondence with the integers, so its transfinite cardinal must be bigger than $\aleph_0$. Indeed, given any transfinite cardinal, there is always a larger one. thus Willink's infinity, determined by the sequence of integers 1, 2, 3, … going on forever, is merely the *smallest* infinity. There is *no* largest one, no "Highest Space" fit for the Almighty.

4 One of the high points of Greek geometry, and the climax of Euclid's Elements, is the discovery that there are exactly five regular solids. By definition, these are solids whose faces are regular polygons, arranged in exactly the same manner at each vertex. They are (see Figure 36):

- *Tetrahedron,* with four equilateral triangles as faces, three meeting at each vertex
- *Cube* (or *hexahedron*), with six squares as faces, three meeting at each vertex

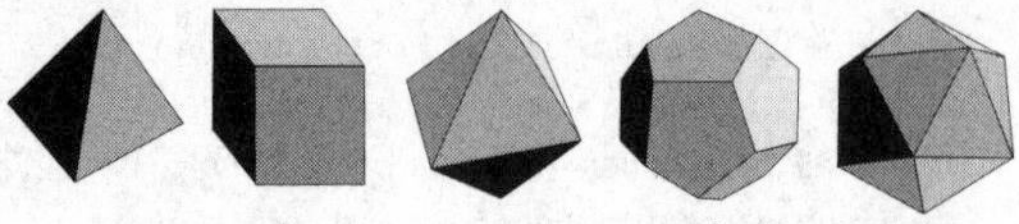

Figure 36 The five regular solids in three dimensions.
(**a**) Tetrahedron. (**b**) Cube. (**c**) Octahedron. (**d**) Dodecahedron.
(**e**) Icosahedron.

- *Octahedron,* with eight equilateral triangles as faces, four meeting at each vertex
- *Dodecahedron,* with twelve equilateral pentagons as faces, three meeting at each vertex
- *Icosahedron,* with twenty equilateral triangles as faces, five meeting at each vertex

The final book (Book XIII) of the *Elements* is aimed at proving (by construction) the existence of the five regular solids, and the other twelve books supply the necessary logical buildup. The tetrahedron is constructed in Proposition 13 of Book XIII, the octahedron in Proposition 14, the cube in Proposition 15, the icosahedron in Proposition 16, and the dodecahedron in Proposition 17. Proposition 18 is "To set out the sides of the five figures and to compare them with one another." Here all five are assumed to be inscribed in the same sphere. This is followed by an unnumbered statement: "No other figure, beside the said five figures, can be constructed which is contained by equilateral and equiangular figures equal to one another." That is, there are exactly five regular solids. This statement is proved, there is one further unnumbered lemma (an auxiliary proposition) justifying one technical point in the proof, and the book — and with it the entire opus — ends. The classification and construction of the five regular solids are thus the climax of the *Elements* and may well have been its main objective.

5 Abbott now begins to unfold his central scientific message, and A. Square begins his first tentative steps along the path that human mathematicians followed in the nineteenth and twentieth centuries.

thirst, for more knowledge. Doubtless we cannot *see* that other higher Spaceland now, because we have no eye in our stomachs. But, just as there *was* the realm of Flatland, though that poor puny Lineland Monarch could neither turn to left nor right to discern it, and just as there *was* close at hand, and touching my frame, the land of Three Dimensions, though I, blind senseless wretch, had no power to touch it, no eye in my interior to discern it, so of a surety there is a Fourth Dimension, which my Lord perceives with the inner eye of thought. And that it must exist my Lord himself has taught me. Or can he have forgotten what he himself imparted to his servant?

In One Dimension, did not a moving Point produce a Line with *two* terminal points?

In Two Dimensions, did not a moving Line produce a Square with *four* terminal points?

In Three Dimensions, did not a moving Square produce — did not this eye of mine behold it — that blessed Being, a Cube, with *eight* terminal points?

And in Four Dimensions shall not a moving Cube — alas, for Analogy, and alas for the Progress of Truth, if it be not so — shall not, I say, the motion of a divine Cube result in a still more divine Organization[8] with *sixteen* terminal points?

Behold the infallible confirmation[9] of the Series, 2, 4, 8, 16: is not this a Geometrical Progression? Is not this — if I might quote my Lord's own words — "strictly according to Analogy"?

Again, was I not taught by my Lord that as in a Line there are *two* bounding Points, and in a Square there are *four* bounding Lines, so in a Cube there must be *six* bounding Squares? Behold once more the confirming Series, 2, 4, 6: is not this an Arithmetical Progression? And consequently does it not of necessity follow that the more divine offspring of the divine Cube in the Land of Four Dimensions, must have 8 bounding Cubes: and is not this also, as my Lord has taught me to believe, "strictly according to Analogy"?

O, my Lord, my Lord, behold, I cast myself in faith upon conjecture, not knowing the facts; and I appeal to your Lordship to confirm or deny my logical anticipations. If I am wrong, I yield, and will no longer demand a fourth Dimension; but, if I am right, my Lord will listen to reason.

I ask therefore, is it, or is it not, the fact, that ere now your countrymen also have witnessed the descent of Beings of a higher order than their own, entering closed rooms, even as your Lordship entered mine, without the opening of doors or windows, and appearing and vanishing at will?[10] On the reply to this question I am ready to stake everything. Deny it, and I am henceforth silent. Only vouchsafe an answer.

Sphere (after a pause). It is reported so. But men are divided in opinion as to the facts. And even granting the facts, they explain them in different ways. And in any case, however great may be the number of different explanations, no one has

The concept was first given voice by Gauss in his 1827 *Treatise on the Geometry of Curved Surfaces,* where he invoked the image of intelligent flatworms moving along a curved membrane. Sartorius von Waltershausen's 1860 *Biography of Carl Friedrich Gauss* states,

> This great man was used to say [that is, customarily said] that, as we can conceive beings (like infinitely attenuated bookworms in an infinitely thin sheet of paper) which possess only the notion of space of two dimensions, so we may imagine beings capable of realising space of four or a greater number of dimensions.

Sylvester, interviewing von Waltershausen in 1869, reported the same statement.

In 1846 the psychologist Gustav Theodor Fechner (1801–1887) wrote "Why Space Has Four Dimensions," part of his collection *Vier Paradoxe* (*Four Paradoxes*) published under the pseudonym of Dr. Mises. He employed the same "dimensional analogy" that Abbott later used: Imagine a two-dimensional creature trying to comprehend our three-dimensional world; then beef everything up a dimension. Fechner's two-dimensional creature was a shadow-man projected onto a vertical screen. Fechner suggested that a creature from the third dimension might be able to demonstrate its existence to the shadow-man: "I take the plane upon which my shadow-man is, and move it through three dimensions. Thus the shadow-man perceives this third dimension. The man himself may be changed, and, at the end of his trip be pale and rumpled, when he had started the trip cosy and flat." However, this argument is not entirely convincing. If the Earth were moved along a fourth dimension, would we notice? How do we know it's not happening right *now*?

Fechner's story was reprinted in his *Kleine Schriften* (*Short Writings*) of 1875, and in 1876 the British journal *Mind* referred to his work when J. M. P. Land wrote a reply to an article by Fechner's colleague Helmholtz. Helmholtz's work had been made available in English translation in 1870, in the journal

Academy, and the article in *Mind* was a further extension. Remember that it was Helmholtz who liked to explain non-Euclidean geometry in terms of a creature that tries to discover the geometry of its universe by purely intrinsic means, and Abbott may have heard about this idea from John Tyndall when they both visited George Eliot. Eliot lived as wife (in all but marriage) with George Henry Lewes (1817–1878), whose son Charles Lee Lewes was also present at the dinner with Tyndall. Lewes was interested in Kant's ideas about higher-dimensional geometry. He was also stimulated by Sylvester's 1869 speech "A Plea for the Mathematician" (*Nature* 1 [1969–70] 289; reprint of "Address to the Mathematics and Physics Section," *Report of the 39th Meeting of the British Association for the Advancement of Science,* Exeter 1869, published by John Murray, London, 1870, 1–9), which asserted the value of geometric generalizations in higher dimensions as well as merely algebraic ones.

The main thrust of "A Plea for the Mathematician" was an eloquent case for the abandonment of Euclid's *Elements* and its derivatives as texts for school geometry — a frequent topic of debate in the 1860s, which met with considerable resistance from educational conservatives. In 1870 a meeting of thirty-six public school headmasters voted overwhelmingly in favor of a resolution to reevaluate Euclid as a text; in 1871 the British Association set up, to do just that, a special committee that included Cayley, Clifford, Thomas Archer Hirst (1830–1892), Salmon, Henry J. S. Smith (1826–1883), and Sylvester. A pressure group, the Association for the Improvement of Geometrical Teaching, was also set up in that year, and by 1875 membership was approaching a hundred, and a new geometry syllabus had been agreed upon and published. Abbott's opposite number at Cheltenham Ladies' College, Dorothea Beale, became the first woman member in the same year. The association generated some rather feeble substitute texts, and it produced reports until 1893, but it failed to bring about any substantial changes.

adopted or suggested the theory of a Fourth Dimension. Therefore, pray have done with this trifling, and let us return to business.

I. I was certain of it. I was certain that my anticipations would be fulfilled. And now have patience with me and answer me yet one more question, best of Teachers! Those who have thus appeared—no one knows whence—and have returned—no one knows whither—have they also contracted their sections and vanished somehow into that more Spacious Space,[11] whither I now entreat you to conduct me?

Sphere (*moodily*). They have vanished, certainly —if they ever appeared. But most people say that these visions arose from the thought—you will not understand me—from the brain; from the perturbed angularity of the Seer.

I. Say they so? Oh, believe them not. Or if it indeed be so, that this other Space is really Thoughtland, then take me to that blessed Region where I in Thought shall see the insides of all solid things. There, before my ravished eye, a Cube, moving in some altogether new direction, but strictly according to Analogy, so as to make every particle of his interior pass through a new kind of Space, with a wake of its own—shall create a still more perfect perfection than himself, with sixteen terminal Extra-solid angles, and Eight solid Cubes for his Perimeter. And once there, shall we stay our upward course? In that blessed region of Four Dimensions, shall we linger on the threshold of the

Fifth,[12] and not enter therein? Ah, no! Let us rather resolve that our ambition shall soar with our corporal ascent. Then, yielding to our intellectual onset, the gates of the Sixth Dimension shall fly open; after that a Seventh, and then an Eighth——

How long I should have continued I know not. In vain did the Sphere, in his voice of thunder, re-iterate his command of silence, and threaten me with the direst penalties if I persisted. Nothing could stem the flood of my ecstatic aspirations. Perhaps I was to blame; but indeed I was intoxicated with the recent draughts of Truth to which he himself had introduced me. However, the end was not long in coming. My words were cut short by a crash outside, and a simultaneous crash inside me, which impelled me through space with a velocity that precluded speech. Down! down! down! I was rapidly descending; and I knew that return to Flatland was my doom. One glimpse, one last and never-to-be-forgotten glimpse I had of that dull level wilderness—which was now to become my Universe again—spread out before my eye. Then a darkness. Then a final, all-consummating thunder-peal; and, when I came to myself, I was once more a common creeping Square, in my Study at home, listening to the Peace-Cry of my approaching Wife.

6 This begins the second of the two new passages that Abbott mentions in his preface to the second edition. The insertion continues until "*Sphere.* Analogy! Nonsense: what analogy?" on page 169.

7 Here Abbott neatly turns the tables on those of his critics who argued that Flatland's objects must actually extend a tiny way into the third dimension, by insisting that the analogous statement should be valid one dimension higher. Presumably, the difficult passage on page 141 was included in order to set up the present one.

8 Here A. Square is describing the *hypercube* or *tesseract,* the four-dimensional analogue of the cube. The name tesseract seems to have been coined by C. H. Hinton in his 1902 *The Recognition of the Fourth Dimension.* It also appears in the 1904 "The Fourth Dimension," though in an earlier essay ("What Is the Fourth Dimension?"), he employs the name *four-square.* A variant spelling is tessaract, and according to Henry Parker Manning's 1914 *Geometry of Four Dimensions,* the name "… belongs to a terminology in which the name of a figure designates the number of its axes: pentact, a figure with five axes, penta-tessaract, a regular 16-hedroid, T. Proctor Hall, *Am. Jour. Math.,* vol. 15: 179." The Greek word τεσσαρα (tessara) means "four." Compare the Latin *tessera,* "cube": Confusion between the two languages may have led to the two spellings. Today *hypercube* is by far the most common term. Other names in the mathematical literature include the variant spelling *tessaract* (used by Manning), *measure polytope, octahedroid,* and *8-cell*; in French, *octaédroïde*; and in German *Masspolytop, Achtzell,* and *Oktaschem.*

To gain insight into the hypercube, we may continue the table of number patterns in the natural manner (see table, at top of page 174). We deduce that the hypercube should be made from eight cubes joined face-to-face. Abbott argues by analogy; later mathematicians would employ coordi-

Dimension	Name	Number of Vertices	Number of Faces
0	point	1	0
1	line	2	2
2	square	4	4
3	cube	8	6
4	hypercube	16	8

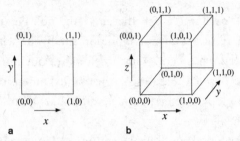

Figure 37 (**a**) Coordinates of the corners of a square. (**b**) Coordinates of the corners of a cube.

nates to give a more precise "construction." The idea is simple and can be described here. In two-dimensional space, a unit square can be specified by the coordinates (x, y) of its four vertices:

$$(0, 0)\,(0, 1)\,(1, 0)\,(1, 1)$$

which are all possible pairs formed from the digits 0, 1 (Figure 37a). Similarly in three-dimensional space, a unit cube can be specified by the coordinates (x, y, z) of its eight vertices:

$$(0, 0, 0)\,(0, 0, 1)\,(0, 1, 0)\,(0, 1, 1)$$
$$(1, 0, 0)\,(1, 0, 1)\,(1, 1, 0)\,(1, 1, 1)$$

which are all possible triples formed from the digits 0, 1 (Figure 37b). It is therefore reasonable to consider all possible quadruples formed from the digits 0, 1 in four-dimensional space, namely the sixteen points (x, y, z, w) given by

$$(0, 0, 0, 0)\,(0, 0, 0, 1)\,(0, 0, 1, 0)\,(0, 0, 1, 1)$$
$$(0, 1, 0, 0)\,(0, 1, 0, 1)\,(0, 1, 1, 0)\,(0, 1, 1, 1)$$
$$(1, 0, 0, 0)\,(1, 0, 0, 1)\,(1, 0, 1, 0)\,(1, 0, 1, 1)$$
$$(1, 1, 0, 0)\,(1, 1, 0, 1)\,(1, 1, 1, 0)\,(1, 1, 1, 1)$$

These define the vertices of the hypercube.

Can we find its eight cubic faces? As a warm-up, consider the analogous problem: Find the six square faces of the cube. To obtain the answer, observe that each face of the cube lies in a plane perpendicular to some coordinate axis and that there are six relevant planes: $x = 0$, $x = 1$, $y = 0$, $y = 1$, $z = 0$, and $z = 1$. That is, we fix the value of one coordinate to be either 0 or 1. For example, if we fix $y = 0$ and

look for the vertices satisfying that condition, we find exactly four of them:

$$(0, 0, 0)\,(0, 0, 1)\,(1, 0, 0)\,(1, 0, 1)$$

If we omit the y-coordinate, the one we fixed at 0, we find that these are just the four vertices $(0, 0)$ $(0, 1)\,(1, 0)\,(1, 1)$ of the square. In this manner, we can find the cube's six square faces algebraically.

For the hypercube, there are eight "hyperplanes" perpendicular to the four axes: $x = 0$, $x = 1$, $y = 0$, $y = 1$, $z = 0$, $z = 1$, $w = 0$, and $w = 1$. For each of these, we obtain a cube of vertices. For example, again choosing $y = 0$, we get the eight points

$$(0, 0, 0, 0)\,(0, 0, 0, 1)\,(0, 0, 1, 0)\,(0, 0, 1, 1)$$
$$(1, 0, 0, 0)\,(1, 0, 0, 1)\,(1, 0, 1, 0)\,(1, 0, 1, 1)$$

and if we omit the y-coordinate, we find precisely the eight vertices of the cube.

There are many ways to visualize a hypercube. One is *projection*. If a cube is projected into two dimensions, it is possible to arrange the projection so that the result is two squares, one inside the other, joined by their corners (Figure 38a). All six square faces of the cube are then visible: the central square, the four trapezia that surround it, and the outer square. Analogously, a hypercube can be projected into three dimensions (and then drawn in two as usual) to produce two cubes, one inside the other, joined by corresponding vertices and with suitable flat faces attached (Figure 38b). Now, with a little

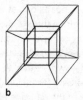

Figure 38 (**a**) Projection of a cube into the plane. (**b**) Analogous projection of a hypercube into three-dimensional space.

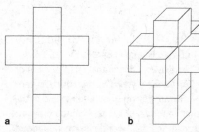

Figure 40 (**a**) Cut out, fold, and make a cube. (**b**) Perform the analogous process with this, and you get a hypercube.

care, all eight component cubes can be seen: the central cube, the six truncated pyramids attached to it, and the outer cube.

A second way is to *slice* the hypercube into a family of three-dimensional sections, the method favored by Hinton and Alicia Stott Boole. If the slices are parallel to a cubic face, then the result is either empty or a cube. However, if the slice is at right angles to the main diagonal of the hypercube, the resulting sequence of solids is more complicated (Figure 39). Such a sequence, in the form of a series of sculptures, used to exist (and for all I know still does) beside the railway line from Kiel to Hamburg, in Germany.

A third method is to *unfold*. A cube made of cardboard can be cut along some of its edges and unfolded into a cross shape formed by six squares (Figure 40a). In the same way, a hypercube can be unfolded into a cross shape formed by eight cubes (Figure 40b). Salvador Dali used such an image in his 1954 painting *Crucifixion* (otherwise known as *Corpus Hypercubus*), which shows Christ crucified on a hypercubic cross. A four-dimensional building in the form of a hypercube forms the basis of the 1941 SF story "And He Built a Crooked House" by Robert A. Heinlein (1907–1988). Gardner's recre-

ational mathematics book *Mathematical Carnival* (1969) contains a chapter on hypercubes, as does Dewdney's *The Armchair Universe* (1988).

9 In the passage that follows, Abbott makes his pitch for his central message, the existence (in some sense) of spaces of any number of dimensions. Of course, the mere existence of certain extrapolated numerical patterns cannot of itself prove that something exists to fit those patterns — indeed, there are many apparent numerical patterns in mathematics that fail to extrapolate as expected. For example, the first five Fermat numbers are prime, but not the sixth; and there are "hypercomplex number systems" of dimensions 1, 2, 4, and 8 — but not 16 or any higher power of 2. The sequence 1, 3, 9, 27, 81 looks like the powers of 3 and therefore continues with 729, except that if in fact the sequence lists those n that divide $2^n + 1$, it actually continues with 171. There are endless examples of a similar kind. Despite these objections, the construction of n-space by n-tuples of real numbers is too technical for Abbott's tale, and its justification is too abstruse: The numerical patterns themselves are very compelling and can be justified by other means, so he wisely restricts his tale to those.

10 In our universe, no such appearance or disappearance has ever been demonstrated scientifically, but many reports can be found in the works of American journalist Charles Hoy Fort (1874–1932): *The Book of the Damned* (1919), *New Lands*

Figure 39 Slices of a hypercube perpendicular to its main diagonal.

(1923), *Lo!* (1931), and *Wild Talents* (1932). For example, Ambrose Bierce (1842–1914?), best known for *The Devil's Dictionary* (1911, variant of the earlier *The Cynic's Word Book,* 1906), disappeared from Texas, having gone into Mexico at the height of that country's civil war. At much the same time, one Ambrose Small went missing in Canada. Fort wondered (not seriously: his books display an excellent sense of dry humor) whether someone was collecting Ambroses.... The Fortean Society, founded in 1931, continues his work and evinces a similar sense of humor.

There are many variants on the ability of creatures from higher dimensions to access the interiors of closed regions of lower dimensions. In *The Fourth Dimension,* Rudy Rucker points out that "a four-dimensional creature could drink up your Chivas Regal without ever opening the bottle!" In 1827 the mathematician Augustus Ferdinand Möbius (1790–1868) realized that the silhouette of a left hand can be rotated into that of a right hand by using a third dimension and also suggested that if a fourth dimension were available, the same could be done with a 3D hand. H. G. Wells made use of this idea in his 1897 "The Plattner Story." A teacher in a boy's school (!) is thrust into the fourth dimension when a green substance found by his pupils explodes. His new world is inhabited by ghost-like creatures, four-dimensional extensions of normal three-dimensional people. When Plattner returns to three-dimensional space, his body has been left-right reflected.

Philosophers and pseudoscientists have seized on such phenomena and used them to explain such things as ghosts and the spirit world — and, on occasion, to "prove" their existence. Möbius is best known for his *Möbius band* (or *strip*), a single-sided surface formed by joining a strip of paper end to end with a half-twist; see *Möbius and His Band* (edited by Raymond Flood, John Fauvel, and Robin Wilson).

A notoriously naive believer in the reality of a fourth dimension was Johann Carl Friederich Zöllner (1834–1882), astronomy professor at Leipzig — the

institution, Rucker points out, at which Möbius discovered how to turn a three-dimensional object into its mirror image by using a rotation in four dimensions, and where Fechner's 1846 essay "Why Space Has Four Dimensions" was written. In 1875 Zöllner visited William (later Sir William) Crookes (1832–1919), English inventor of the cathode ray tube and discoverer of the element thallium, and became interested in spiritualism. Crookes's enthusiasm was based on demonstrations performed by the American medium Henry Slade. Slade made the concept of the fourth dimension notorious when, in 1877, on a visit to England during which he held seances with prominent people, he was put on trial for fraud and charged with "using subtle crafts and devices, by palmistry and otherwise," that is, conjuring tricks. What made the trial sensational was that several prominent scientists, most associated with the Society for Psychical Research, sprang to Slade's defense. Zöllner, who was looking for proof of the existence of a four-dimensional spirit world and was convinced that Slade had provided it, was especially vocal. Other scientists involved — all physicists, as it happened — included Crookes; Wilhelm Eduard Weber (1804–1891), who collaborated with Gauss in his work on magnetism; Sir J(oseph) J(ohn) Thomson (1856–1940), who won a Nobel Prize in 1906 for discovering the electron; and Lord Rayleigh (John William Strutt, Third Baron, 1842–1919), a Nobel winner (in 1904) for isolating the inert gas argon.

Zöllner and his physicist colleagues were convinced by demonstrations of the following kind. Slade started with an unknotted loop whose ends were attached to each other and sealed and then miraculously tied four knots in it. In three dimensions there is no legitimate way to achieve this feat, but it would be trivial in four (see chapter 4 of *Flatterland,* "A Hundred and One Dimensions"). It is also a simple conjuring trick in three dimensions, but poor Zöllner was too excited to think of that. (Modern scientists have been similarly taken in by alleged paranormal phenomena that are child's play to any member of the Magic Circle. Physicists seem especially prone to falling into this trap, probably because their training is based on making far-reaching and counterintuitive deductions from the evidence of their senses but does not condition them to guard against deliberate deception.) In *The Unexpected Hanging and Other Mathematical Diversions* (1969), Martin Gardner explains Slade's probable methods for performing two other demonstrations. One starts with a strip of leather in which two parallel cuts have been made (Figure 41a), and under the table the strip becomes braided (Figure 41b). The secret of this startling manipulation is purely topological and well known to leather workers and Boy Scouts. It is explained in many books, including the classic *The Ashley Book of Knots* (1944), where it is knot 2954 on page 486. Nearby are numerous other conjuring tricks with knots. Figure 41c shows how it is done. In the second trick, an elastic band inside a sealed box becomes knotted. Gardner calls Slade "one of the most colourful and successful mountebanks in the history of American spiritualism."

Zöllner described the outcome of, but not the trickery behind, many other such experiments in his *Transcendental Physics* of 1879, which was translated into English in 1881. These included turning snail shells into their mirror images and similarly inverting the molecules of the chemical dextrotartaric acid to obtain levotartaric acid, which rotates polar-

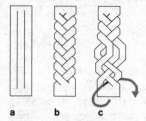

Figure 41 A strip of leather with cuts (**a**) becomes braided (**b**). Manipulations in the fourth dimension? Not at all (**c**). Braid the top half of the strip: the lower half automatically braids in the opposite direction. Then work the lower half as shown by the arrow.

ized light in the opposite direction. A close reading of the text reveals that these experiments did not actually *work,* and a little thought makes it clear that whereas they would be easy for someone with genuine access to the fourth dimension, they pose serious problems for a conjuror (how to get hold of the two forms of tartaric acid, for instance), but Zöllner remained undeterred. To quote Rucker again,

> The main effect of Zöllner's work was that the fourth dimension began to seem disreputable and unscientific. Yet his basic message, *spirits live in the fourth dimension,* did not fall on totally deaf ears. The notion of beings living in an unseen hyperspace world was taken up by turn-of-the-century Protestant clergymen all over England. The best known of these clergymen is, of course, Edwin Abbott, but there were many others who enthusiastically adopted the notion that heaven, hell, our souls, the angels, and God himself could be comfortably lodged in some higher dimension.

H. G. Wells made use of such beliefs in "The Wonderful Visit" (1895), in which an angel appears in our universe, having come from a four-dimensional space with no gravity. The Angel, discussing dreams with the Vicar, comes to an astonishing realization:

> "And in some incomprehensible manner I have fallen into this world of yours out of my own!" said the Angel, "into the world of my dreams grown real." He looked about him. "Into the world of my dreams."
>
> "It is confusing," said the Vicar. "It almost makes one think there may be (ahem) Four Dimensions after all. In which case, of course," he went on hurriedly — for he loved geometrical speculations and took a certain pride in his knowledge of them — "there may be any number of three-dimensional universes packed side by side, and all dimly dreaming of one another. There may be world upon world, universe upon universe."

This is an early occurrence of the SF idea of parallel worlds. Abbott, it must be said, was generally unimpressed by the use of four-dimensional concepts to "explain" spiritual matters. In *The Kernel and the Husk* (1887) he says,

> You know — or might know if you would read a little book recently published called *Flatland,* and still better, if you would study a very able and original work by Mr. C. H. Hinton — that a being of Four Dimensions, if such there were, could come into our closed rooms without opening door or window, nay could even penetrate into, and inhabit our bodies.... Even if we could conceive of Space of Four Dimensions — which we cannot do although we can perhaps describe what some of its phenomena would be if it existed — we should not be one whit better morally or spiritually. It seems to me rather a moral than an intellectual process, to approximate to the conception of a spirit: and toward this no knowledge of Quadridimensional space can guide us.

11 Concepts very close to the existence of additional nonspatial dimensions and to time being a fourth dimension are not solely modern. They can be found in the writings of Saint Augustine (354–430), bishop of Hippo in Roman Africa and the most prominent of the early Christian theologians. Augustine was trying to reconcile the act of creation, which changed the physical universe at a specific instant of time, with a God who was outside time and in eternity and therefore perceived no change. In chapter XI of his *Confessiones* (*Confessions,* c. 400) Augustine says, "Nor was it in the universe that you made the universe, because until the universe was made there was no place where it could be made." Attributing the creation of heaven and earth to God's Word alone, he adds, "You must have created it by an utterance outside time, so that you could use it as a mouthpiece for your decree, uttered in time, that heaven and Earth should be made." Augustine also keeps a clear head on the question of the beginning of time, which I recommend to people who ask what happened before the Big Bang: "if there was no time before heaven and earth were created, how can anyone ask what you [God] were doing 'then'? If there was no time, there was no 'then.'" A few pages later we find this extraordinary passage: "Another view might be that past and future do exist, but that time emerges from some secret refuge when it passes from the future to the present, and goes back into hiding when it moves from the present to the past." And shortly thereafter: "I see time … as an extension of some sort." In chapter XI, section 6 of *De Civitate Dei* (*City of God,* 413–426), he says, "there can be no doubt that the world was not created *in* time but *with* time." That is, space and time came into being together. In section 7 of the same chapter, he says, "[God] does not look ahead to the future, look directly at the present, look back to the past. He sees in some other manner, utterly remote from anything we experience or could imagine.… He sees all without any kind of change."

12 Having grasped Abbott's central analogy, as explained to him by the Sphere, A. Square finds the logic of the argument so compelling that he extracts more from it than the Sphere intended. The sequence of numbers 0, 1, 2, 3, 4, 5, … extends forever, without limit (for if n were the largest whole number, what would $n + 1$ be?). So do the Sphere's combinatorial sequences 1, 2, 4, 8, … and 0, 2, 4, 6, … Why, then, stop at four dimensions? The same thought occurred to the mathematicians who first investigated four-dimensional geometry (see "The Fourth Dimension in Mathematics" in this book), and their answer was the same as A. Square's. Indeed, the number patterns alluded to by the Sphere can be extended to n dimensions. Here the number of n-tuples of digits 0, 1 is 2^n, and these n-tuples define the vertices of the n-cube. There is also an n-dimensional analogue of the octahedron, known as a *cross polytope,* formed by the midpoints of the $(n - 1)$-dimensional "faces" of the n-cube, and an n-dimensional analogue of the tetrahedron, called an *n-simplex.* For n greater than or equal to 5, these are the *only* regular hypersolids (the technical term is *polytopes*). The dodecahedron and the icosahedron are special to three dimensions. In four dimensions, Alicia Boole and Schoute found three new "regular hypersolids" called the *24-cell* (with 24 octahedral faces), the *120-cell* (120 icosahedral faces), and the *600-cell* (600 dodecahedral faces). Figure 42 shows projections of all six regular polytopes.

The history of four-dimensional geometry is outlined in the introduction to Manning's *Geometry of Four Dimensions*. Schoute's results appeared in his *Mehrdimensional Geometrie* (*Multi-dimensional Geometry*) of 1902 and 1905. Another introduction to the area was the 1903 *Géométrie à Quatre Dimensions* (*Four-Dimensional Geometry*) of E. P. Jouffret. Later, it turned out that the Swiss mathematician Ludwig Schläfli (1814–1895) had already discovered these hypersolids between 1850 and 1852, when he wrote his *Theorie der vielfachen Kontinuität* (*Theory of Many-fold Continuity*), but he was unable to find a publisher ("apparently on account of its length," says Manning), and it remained in manuscript form among his papers until it was published in 1911, well after Schläfli's death. Terminology in the area was varied: Greek numerology was common, but straightforward names of the general type "*m*-cell" and "*m*-hedroid" (*m* = 5, 8, 16, 24, 120, 600) were also employed. The table on page 180 shows the technical names used by various authors. Manning's spelling is archaic, and his names for the last four polytopes would now be spelled with *c* in place of *k*. Symbolism was also not standardized: Jouffret used C^5, C^8, C^{16}, C^{24}, C^{120}, C^{600}; Schoute's notation was Z_5, Z_8, Z_{16}, Z_{24}, Z_{120}, Z_{600}; and Schläfli's symbols, which indicated the component polyhedral "faces" and their arrangement, were {3, 3, 3}, {3, 3, 4}, {4, 3, 3}, {3, 4, 3}, {5, 3, 3}, and {3, 3, 5}; see Coxeter's *Regular Polytopes*.

The two-dimensional case is also exceptional: Here there are infinitely many regular "solids" — namely, the regular polygons. And in one dimension

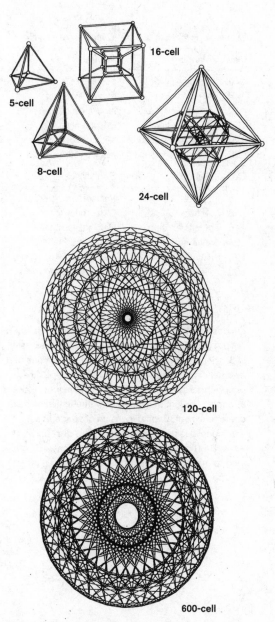

Figure 42 The six regular polytopes in four dimensions, here shown as projections into three-dimensional space.

Terminology for Regular Four-Dimensional Polytopes

Cells	Manning	Jouffret
5	pentahedroid	pentaédroïde
8	octahedroid	octaédroïde
16	hexadekahedroid	hexadecaédroïde
24	ikosatetrahedroid	icosatétraédroïde
120	hekatonikosahedroid	hécatonicosaédroïde
600	hexakosioihedroid	hexacosiédroïde

Cells	Schoute	Schläfli
5	Fünfzell	Pentaschem
8	Achtzell	Oktaschem
16	Sechszehnzell	Hekkaidekaschem
24	Vierundzwanzigzell	Eikositetraschem
120	Hundertzwanzigzell	Hekatonkaieikosaschem
600	Sechshundertzell	Hexakosioschem

the only regular "solid" is a line segment. Hence in
dimensions 1, 2, 3, 4 the number of regular "solids"
is 1, ∞ (infinity), 5, 6; and in all higher dimensions it
is 3. These numbers show that not all properties of
n-dimensional space can be found by simple analogies with space of two and three dimensions. More
strongly, they hint at a fact that mathematicians have
come to expect: Each dimension of space has its
own special features, unlike any other dimension.
As well as having many other features that are common to most or all dimensions, of course.

§ 20. — How the Sphere encouraged me in a Vision.

ALTHOUGH I had less than a minute for reflection, I felt, by a kind of instinct, that I must conceal my experiences from my Wife. Not that I apprehended, at the moment, any danger from her divulging my secret, but I knew that to any Woman in Flatland the narrative of my adventures must needs be unintelligible. So I endeavoured to reassure her by some story, invented for the occasion, that I had accidentally fallen through the trapdoor of the cellar, and had there lain stunned.

The Southward attraction in our country is so slight that even to a Woman my tale necessarily appeared extraordinary and well-nigh incredible; but my Wife, whose good sense far exceeds that of the average of her Sex, and who perceived that I was unusually excited, did not argue with me on the subject, but insisted that I was ill and required repose. I was glad of an excuse for retiring to my chamber to think quietly over what had happened. When I was at last by myself, a drowsy sensation fell on me; but before my eyes closed I endeavoured to reproduce the Third Dimension, and especially the process by which a Cube is con-

1 Abbott hedges his bets by making the Sphere acknowledge the correctness of A. Square's arguments (on the existence of arbitrarily many dimensions) by setting it within a dream. The dream occurs in a fictitious setting and is thus *doubly fictitious*. In Victorian times, casting a story as a dream was a literary cliché, and more than one otherwise compelling story ends with the reader being informed that "it was all a dream." This is the case, for instance, with *Alice's Adventures in Wonderland:*

> At this the whole pack [of playing cards] rose up in the air, and came flying down upon her; she gave a little scream, half of fright and half of anger, and tried to beat them off, and found herself lying on the bank, with her head in the lap of her sister, who was gently brushing away some dead leaves that had fluttered down from the trees upon her face.
>
> "Wake up, Alice dear!" said her sister. "Why, what a long sleep you've had!"
>
> "Oh, I've had such a curious dream!" said Alice.

Here, the statement that the story was really just a dream does not spoil the illusion — perhaps because a significant feature of *Alice* is its remarkably dream-like quality — but modern readers usually find such revelations anticlimactic and irritating.

2 Abbott rounds out his analogy by reducing the dimensions of space even below the limited realm of Lineland. (Recall that Poincaré and Menger went one stage further and assigned dimension *minus one* to the empty set, but it is not surprising that Abbott did not go so far.) The Being that inhabits Pointland is the ultimate solipsist, and Abbott can make good use of him to castigate complacency and lack of imagination. The wonderful thing about solipsists is that you can insult them as much as you want and they can only blame it on themselves.

structed through the motion of a Square. It was not so clear as I could have wished; but I remembered that it must be "Upward, and yet not Northward," and I determined steadfastly to retain these words as the clue which, if firmly grasped, could not fail to guide me to the solution. So mechanically repeating, like a charm, the words, "Upward, yet not Northward," I fell into a sound refreshing sleep.

During my slumber I had a dream.[1] I thought I was once more by the side of the Sphere, whose lustrous hue betokened that he had exchanged his wrath against me for perfect placability. We were moving together towards a bright but infinitesimally small Point, to which my Master directed my attention. As we approached, methought there issued from it a slight humming noise as from one of your Spaceland blue-bottles, only less resonant by far, so slight indeed that even in the perfect stillness of the Vacuum through which we soared, the sound reached not our ears till we checked our flight at a distance from it of something under twenty human diagonals.

"Look yonder," said my Guide, "in Flatland thou hast lived; of Lineland thou hast received a vision; thou hast soared with me to the heights of Spaceland; now, in order to complete the range of thy experience, I conduct thee downward to the lowest depth of existence, even to the realm of Pointland, the Abyss of No dimensions.[2]

"Behold yon miserable creature. That Point is a Being like ourselves, but confined to the non-dimensional Gulf. He is himself his own World, his own Universe; of any other than himself he can form no conception; he knows not Length, nor Breadth, nor Height, for he has had no experience of them; he has no cognizance even of the number Two; nor has he a thought of Plurality; for he is himself his One and All, being really Nothing. Yet mark his perfect self-contentment, and hence learn this lesson, that to be self-contented is to be vile and ignorant, and that to aspire is better than to be blindly and impotently happy. Now listen."

He ceased; and there arose from the little buzzing creature a tiny, low, monotonous, but distinct tinkling, as from one of your Spaceland phonographs, from which I caught these words, "Infinite beatitude of existence! It is; and there is none else beside It."

"What," said I, "does the puny creature mean by 'it'?" "He means himself," said the Sphere: "have you not noticed before now, that babies and babyish people who cannot distinguish themselves from the world, speak of themselves in the Third Person? But hush!"

"It fills all Space," continued the little soliloquizing Creature, "and what It fills, It is. What It thinks, that It utters; and what It utters, that It hears; and It itself is Thinker, Utterer, Hearer, Thought, Word, Audition; it is the One, and yet the All in All. Ah, the happiness, ah, the happiness of Being!"

"Can you not startle the little thing out of its complacency?" said I. "Tell it what it really is, as you told me; reveal to it the narrow limitations of Pointland, and lead it up to something higher." "That is no easy task," said my Master; "try you."

Hereon, raising my voice to the uttermost, I addressed the Point as follows:

"Silence, silence, contemptible Creature. You call yourself the All in All, but you are the Nothing: your so-called Universe is a mere speck in a Line, and a Line is a mere shadow as compared with—" "Hush, hush, you have said enough," interrupted the Sphere, "now listen, and mark the effect of your harangue on the King of Pointland."

The lustre of the Monarch, who beamed more brightly than ever upon hearing my words, shewed clearly that he retained his complacency; and I had hardly ceased when he took up his strain again. "Ah, the joy, ah, the joy of Thought! What can It not achieve by thinking! Its own Thought coming to Itself, suggestive of Its disparagement, thereby to enhance Its happiness! Sweet rebellion stirred up to result in triumph! Ah, the divine creative power of the All in One! Ah, the joy, the joy of Being!"

"You see," said my Teacher, "how little your words have done. So far as the Monarch understands them at all, he accepts them as his own—for he cannot conceive of any other except himself—and plumes himself upon the variety of 'Its Thought' as an instance of creative Power. Let us leave this God of Pointland to the ignorant fruition of his omnipresence and omniscience: nothing that you or I can do can rescue him from his self-satisfaction."

After this, as we floated gently back to Flatland, I could hear the mild voice of my Companion pointing the moral of my vision, and stimulating me to aspire,[3] and to teach others to aspire. He had been angered at first — he confessed — by my ambition to soar to Dimensions above the Third; but, since then, he had received fresh insight, and he was not too proud to acknowledge his error to a Pupil. Then he proceeded to initiate me into mysteries yet higher than those I had witnessed, shewing me how to construct Extra-Solids[4] by the motion of Solids, and Double Extra-Solids by the motion of Extra-Solids, and all "strictly according to Analogy," all by methods so simple, so easy, as to be patent even to the Female Sex.

3 Abbott was a great teacher. Many tributes from former pupils are reported in *City of London School,* and they are made credible by Abbott's general reputation — that is, they are not just polite remarks made at Old Boys' dinners with the benefit of rosy memories and a surfeit of wine. For instance, the Rt. Hon. H. H. Asquith, who became prime minister in 1908, said that Abbott "taught his pupils as well as any schoolmaster of his time." According to William Bramwell Booth (1856–1929), second general of the Salvation Army, "The chief service rendered me … came through the encouragement of Dr. Abbott, the brilliant Head.… He kindled a little flame of intellectual aspiration." And Oscar Browning, a prominent university academic, made his views crystal clear: "When I have been asked who was the most distinguished headmaster in England I have never had any hesitation in replying 'Dr. Abbott'." Characteristically, Abbott draws a more positive lesson from the self-satisfied Being of Pointland than merely berating the unimaginative for their complacency.

4 Today's mathematics has taken Abbott's message to heart, although it followed its own route to get there. An important example of the use of high-dimensional geometry occurs in celestial mechanics, the motion of stars and planets. In this subject, the bodies concerned are usually reduced to point masses, and the state of each such body is represented by *six* numbers: three coordinates of position and a further three coordinates of velocity. (One of the features of Newton's Laws of Motion is that the future trajectory of a particle is determined not by its position alone but also by its velocity. Think of throwing a ball, and you'll see why the velocity matters.) A typical problem in celestial mechanics is the motion of a three-body system, such as the sun, moon, and Earth. This problem involves eighteen independent "coordinates" — six for the sun, six more for the moon, and another six for the Earth. The "phase space" of all possible states of the three-body system therefore has *eighteen* dimensions.

The nine planets of the solar system, plus the sun, form a *sixty*-dimensional mathematical system. At a conservative estimate, the human body, with its numerous flexible joints, is 101-dimensional. And mathematicians have found that it really pays to set up the problem that way and to use geometric concepts (all originally derived by analogy with two and three dimensions) to help solve it.

Multidimensional spaces are important in other areas of human activity as well; a good example is economics. A national economy depends on the sales and purchases of, say, a million different goods. The price of each good is in principle an independent quantity, so the state of the economy can be represented by a "price vector" in a *million*-dimensional space. Again, mathematicians have found that it pays to set up the problem that way and to use geometric concepts to help solve it. An example is the simplex method invented by American mathematician George Bernard Dantzig (1914–) to answer such questions as "How should the economic activity be divided among various goods?" The simplex method is based on an analogy with triangular faces of a three-dimensional polyhedron. (See chapter 4 of *Flatterland,* "A Hundred and One Dimensions.")

§ 21. — How I tried to teach the Theory of Three Dimensions to my Grandson, and with what success.

I AWOKE rejoicing, and began to reflect on the glorious career before me. I would go forth, methought, at once, and evangelize the whole of Flatland. Even to Women and Soldiers should the Gospel of Three Dimensions be proclaimed. I would begin with my Wife.

Just as I had decided on the plan of my operations, I heard the sound of many voices in the street commanding silence. Then followed a louder voice. It was a herald's proclamation. Listening attentively, I recognized the words of the Resolution of the Council, enjoining the arrest, imprisonment, or execution of any one who should pervert the minds of the people by delusions, and by professing to have received revelations from another World.

I reflected. This danger was not to be trifled with. It would be better to avoid it by omitting all mention of my Revelation, and by proceeding on the path of Demonstration—which after all, seemed so simple and so conclusive that nothing would be lost by discarding the former means.

"Upward, not Northward"—was the clue to the whole proof. It had seemed to me fairly clear before I fell asleep; and when I first awoke, fresh from my dream, it had appeared as patent as Arithmetic; but somehow it did not seem to me quite so obvious now. Though my Wife entered the room opportunely just at that moment, I decided, after we had exchanged a few words of commonplace conversation, not to begin with her.

My Pentagonal Sons were men of character and standing, and physicians of no mean reputation, but not great in mathematics, and, in that respect, unfit for my purpose. But it occurred to me that a young and docile Hexagon, with a mathematical turn, would be a most suitable pupil. Why therefore not make my first experiment with my little precocious Grandson, whose casual remarks on the meaning of 3^3 had met with the approval of the Sphere? Discussing the matter with him, a mere boy, I should be in perfect safety; for he would know nothing of the Proclamation of the Council; whereas I could not feel sure that my Sons—so greatly did their patriotism and reverence for the Circles predominate over mere blind affection—might not feel compelled to hand me over to the Prefect, if they found me seriously maintaining the seditious heresy of the Third Dimension.

But the first thing to be done was to satisfy in some way the curiosity of my Wife, who naturally wished to know something of the reasons for which the Circle had desired that mysterious interview, and of the means by which he had entered the house. Without entering into the details of the elaborate account I gave her, — an account, I fear, not quite so consistent with truth as my Readers in Spaceland might desire, — I must be content with saying that I succeeded at last in persuading her to return quietly to her household duties without eliciting from me any reference to the World of Three Dimensions. This done, I immediately sent for my Grandson; for, to confess the truth, I felt that all that I had seen and heard was in some strange way slipping away from me, like the image of a half-grasped, tantalizing dream, and I longed to essay my skill in making a first disciple.

When my Grandson entered the room I carefully secured the door. Then, sitting down by his side and taking our mathematical tablets, — or, as you would call them, Lines — I told him we would resume the lesson of yesterday. I taught him once more how a Point by motion in One Dimension produces a Line, and how a straight Line in Two Dimensions produces a Square. After this, forcing a laugh, I said, "And now, you scamp, you wanted to make me believe that a Square may in the same way by motion 'Upward, not Northward' produce another figure, a sort of extra Square in Three Dimensions. Say that again, you young rascal."

At this moment we heard once more the herald's "O yes! O yes!" outside in the street proclaiming the Resolution of the Council. Young though he was, my Grandson — who was unusu-

1 Poor A. Square is trapped because of an important feature of dynamical systems known as an *invariant subspace*. The likely laws of dynamics in Flatland — namely those of three dimensions restricted to two — imply that any initial state (of position and velocity) that lies entirely within the Flatland plane can never leave that plane. A. Square can no more move things "Upward, not Northward" than we can make a rabbit disappear from our universe by giving it the right push. (Though with the aid of a top hat....)

ally intelligent for his age, and bred up in perfect reverence for the authority of the Circles—took in the situation with an acuteness for which I was quite unprepared. He remained silent till the last words of the Proclamation had died away, and then, bursting into tears, "Dear Grandpapa," he said, "that was only my fun, and of course I meant nothing at all by it; and we did not know anything then about the new Law; and I don't think I said anything about the Third Dimension; and I am sure I did not say one word about 'Upward, not Northward,' for that would be such nonsense, you know. How could a thing move Upward, and not Northward? Upward and not Northward! Even if I were a baby, I could not be so absurd as that. How silly it is! Ha! ha! ha!"

"Not at all silly," said I, losing my temper; "here for example, I take this Square," and, at the word, I grasped a moveable Square, which was lying at hand—"and I move it, you see, not Northward but—yes, I move it Upward—that is to say, not Northward, but I move it somewhere—not exactly like this,[1] but somehow—" Here I brought my sentence to an inane conclusion, shaking the Square about in a purposeless manner, much to the amusement of my Grandson, who burst out laughing louder than ever, and declared that I was not teaching him, but joking with him; and so saying he unlocked the door and ran out of the room. Thus ended my first attempt to convert a pupil to the Gospel of Three Dimensions.

§ 22. — How I then tried to diffuse the Theory of Three Dimensions by other means, and of the result.

MY FAILURE with my Grandson did not encourage me to communicate my secret to others of my household; yet neither was I led by it to despair of success. Only I saw that I must not wholly rely on the catch-phrase, "Upward, not Northward," but must rather endeavour to seek a demonstration by setting before the public a clear view of the whole subject; and for this purpose it seemed necessary to resort to writing.

So I devoted several months in privacy to the composition of a treatise on the mysteries of Three Dimensions. Only, with the view of evading the Law, if possible, I spoke not of a physical Dimension, but of a Thoughtland[1] whence, in theory, a Figure could look down upon Flatland and see simultaneously the insides of all things, and where it was possible that there might be supposed to exist a Figure environed, as it were, with six Squares, and containing eight terminal Points. But in writing this book I found myself sadly hampered by the impossibility of drawing such diagrams as were necessary for my purpose; for of course, in our

1 Abbott now extends the scope of his vision. Not only can higher-dimensional spaces be thought of as new universes, but also there is a wider universe that embodies all conceivable concepts, real or imaginary. The "Mathiverse" of *Flatterland* is an explicit version of the same idea. The big question here is in what sense this "universe" of concepts exists? There are many philosophical paradigms concerning the meaning of mathematical existence. The two most popular are Platonism, which holds that all mathematical concepts preexist in an "ideal" but nonphysical world, and social constructivism, which considers a mathematical concept to be nothing more than a belief that is shared by several different human minds. "Thoughtland," despite its name, is very close to Platonism. In the philosophy of Plato (427–347 B.C.), the human world is merely an image, or projection, of the higher realm of Ideal Forms, as he describes in a famous passage in Book 7 of his *Republic*:

"And do you see," I said, "men passing along the wall, some apparently talking and others silent, carrying vessels and statues and figures of animals made of wood and stone and various materials, which appear over the wall?"

"You have shown me a strange image, and they are strange prisoners."

"Like ourselves," I replied; "and they see only their own shadows, or the shadows of one another, which the fire throws on the opposite wall of the cave?"

"True," he said; "how could they see anything but the shadows if they were never allowed to move their heads?"

"And of the objects which are being carried in like manner they would only see the shadows?"

"Yes," he said.

"And if they were able to talk with one another, would they not suppose that they were naming what was actually before them?"

Mathematicians habitually talk of abstract concepts as though they were real, to the extent of making shapes in the air with their hands while talking about a seven-dimensional sphere; in this they are "closet Platonists," behaving as though their idealized concepts were real objects. This behavior confuses philosophers, who always take everything too seriously, and it makes them think that mathematicians are Platonists and believe that the idealizations *really* exist in some physical sense (but not of *this* universe). On the other hand, there is no serious dispute that mathematics arises through the collective activities of the world's mathematicians, who talk to each other, write their ideas down, and read what their colleagues have written. This behavior confuses a different school of philosophers, who deduce that mathematics (science, too) is merely a shared delusion and could be anything that its practitioners choose it to be — which implies that it is meaningless. (This view has the advantage of relieving said philosophers, and their students, of any need to *understand* mathematics or science.) Strangely, social constructivists always enter rooms through the doorway, rather than choosing to share the collective view that passing through the walls would be equally effective.

The working philosophy of most mathematicians is far more pragmatic and lies somewhere between these extremes, as explained by Reuben Hersh in *What Is Mathematics, Really?* There he says,

> What's the nature of mathematical objects? The question is made difficult by the centuries-old assumption of Western philosophy: "there are two kinds of thing in the world. What isn't physical is mental; what isn't mental is physical." Mental is individual consciousness. It includes private thoughts ... before they're communicated to the world and become social.... Frege [the mathematical logician Gottlob Frege (1848–1925)] showed that mathematical objects are neither physical nor mental. He labelled them "abstract objects." What did he tell us about abstract objects? Only this: they're neither physical nor mental. Are there other things besides numbers that aren't mental or physical? Yes! Sonatas. Prices. Eviction notices. Declara-

country of Flatland, there are no tablets but Lines, and no diagrams but Lines, all in one straight Line and only distinguishable by difference of size and brightness; so that, when I had finished my treatise (which I entitled, "Through Flatland to Thoughtland") I could not feel certain that many would understand my meaning.

Meanwhile my life was under a cloud. All pleasures palled upon me; all sights tantalized and tempted me to outspoken treason, because I could not but compare what I saw in Two Dimensions with what it really was if seen in Three, and could hardly refrain from making my comparisons aloud. I neglected my clients and my own business to give myself to the contemplation of the mysteries which I had once beheld, yet which I could impart to no one, and found daily more difficult to reproduce even before my own mental vision.

One day, about eleven months after my return from Spaceland, I tried to see a Cube with my eye closed, but failed; and though I succeeded afterwards, I was not then quite certain (nor have I been ever afterwards) that I had exactly realized the original. This made me more melancholy than before, and determined me to take some step; yet what, I knew not. I felt that I would have been willing to sacrifice my life for the Cause, if thereby I could have produced conviction. But if I could not convince my Grandson, how could I convince the highest and most developed Circles in the land?

And yet at times my spirit was too strong for me, and I gave vent to dangerous utterances. Already I was considered heterodox if not treasonable, and I was keenly alive to the danger of my position; nevertheless I could not at times refrain from bursting out into suspicious or half-seditious utterances, even among the highest Polygonal and Circular society. When, for example, the question arose about the treatment of those lunatics who said that they had received the power of seeing the insides of things, I would quote the saying of an ancient Circle, who declared that prophets and inspired people are always considered by the majority to be mad; and I could not help occasionally dropping such expressions as "the eye that discerns the interiors of things," and "the all-seeing land"; once or twice I even let fall the forbidden terms "the Third and Fourth Dimensions." At last, to complete a series of minor indiscretions, at a meeting of our Local Speculative Society[2] held at the palace of the Prefect himself,—some extremely silly person having read an elaborate paper exhibiting the precise reasons why Providence has limited the number of Dimensions to Two, and why the attribute of omnividence is assigned to the Supreme alone—I so far forgot myself as to give an exact account of the whole of my voyage with the Sphere into Space, and to the Assembly Hall in our Metropolis, and then to Space again, and of my return home, and of everything that I had seen and heard in fact or vision. At first, in-

tions of war....The US Supreme Court exists. It can condemn you to death! Is the Court physical? If the Court buildings were blown up and the justices moved to the Pentagon, the Court would go on. Is it mental? If all nine justices expired in a suicide cult, they'd be replaced. The Court would go on. The Court isn't the stones of its building, nor is it anyone's minds and bodies.... *It's a social institution.* Mental and physical categories are insufficient to understand it. It's comprehensible only in the context of American society.

Later he adds, "Mathematics consists of concepts. Not pencil or chalk marks, not physical triangles or physical sets, but concepts, which may be suggested or represented by physical objects." The Platonic world of ideals is a useful fiction, providing helpful mental imagery and adding a sense of immediacy and vividness to mathematical research. There is indeed a collective social process, but it is highly constrained by the need for logical consistency with what has gone before: It is not arbitrary.

To Hersh's solution of the nature of mathematics, I'd like to add two things. For "emergent monists" who see mind as a process carried out by a brain, the Cartesian dualism of physical versus mental misses the point. Physical and mental objects are not made of different stuff. *All* objects are physical; *all* processes are carried out using physical objects (or processes carried out by physical objects, or processes carried out by processes carried out by physical objects, or ...). Mental "objects" happen to have two different interpretations: a physical interpretation as a process (this nerve cell fires and sends a signal to that one ...) and a mental one as an experience (the fire feels hot ...). We are easily confused by this interpretation because we experience all physical objects through the medium of our minds. When the mind focuses on *itself*, the scope for confusion is all the greater.

Second, Hersh mentions what looks like a rather good objection to his analysis of the nature of mathematics, made by the science writer Martin Gardner.

When, more than 65 million years ago, two dinosaurs met two other dinosaurs at a water hole, there were then four dinosaurs at the water hole, even though no human was there to think "2 + 2 = 4." Hence 2, 4, and + are not just shared processes in human minds. Hersh offers one answer to this objection, linguistic in nature. I offer another. According to quantum physics, everything in the universe is made from fundamental particles. The classification of four particular collections of fundamental particles as being *the same kind of thing* — dinosaurs — is a human interpretation. The universe would not need to "know" that these subsets are dinosaurs to carry out its rules; all it needs is the quantum physics of particles. Until a human being — or some other entity that classifies reality in the same way — provides an interpretation of the physics, what goes on at the water hole is not a physical instance of "two," or "plus," or "four." It is we who endow the situation with a context in which those concepts apply.

2 Here Abbott is endowing Flatland's inhabitants, and A. Square in particular, with interests similar to those of many Victorians. People gathered regularly in groups to discuss new ideas of science or pseudoscience, or to hear lectures about them, in the belief that they were furthering their education and improving their minds. In general, the Victorians were very keen on the virtues of self-improvement. A. Square, with his strong interests in mathematics and education, would surely have been an enthusiastic participant in Flatland events of this type.

Speculative societies were especially common in Victorian times, when many of the ideas that shaped today's Western world first emerged. There were philosophical societies, societies devoted to the occult, and — especially — societies devoted to spiritualism. Spiritualists believe that the dead can communicate with the living, usually via a medium in a state of trance. Their belief system originated in the United States in 1848, in what Sir Arthur Conan Doyle (1859–1930) later called the Hydesville

deed, I pretended that I was describing the imaginary experiences of a fictitious person; but my enthusiasm soon forced me to throw off all disguise, and finally, in a fervent peroration, I exhorted all my hearers to divest themselves of prejudice and to become believers in the Third Dimension.

Need I say that I was at once arrested and taken before the Council?[3]

Next morning, standing in the very place where but a very few months ago the Sphere had stood in my company, I was allowed to begin and to continue my narration unquestioned and uninterrupted. But from the first I foresaw my fate; for the President, noting that a guard of the better sort of Policemen was in attendance, of angularity little, if at all, under 55°, ordered them to be relieved before I began my defence, by an inferior class of 2° or 3°. I knew only too well what that meant. I was to be executed or imprisoned, and my story was to be kept secret from the world by the simultaneous destruction of the officials who had heard it; and, this being the case, the President desired to substitute the cheaper for the more expensive victims.

After I had concluded my defence, the President, perhaps perceiving that some of the junior Circles had been moved by my evident earnestness, asked me two questions:—

1. Whether I could indicate the direction which I meant when I used the words "Upward, not Northward"?

2. Whether I could by any diagrams or descriptions (other than the enumeration of imaginary sides and angles) indicate the Figure I was pleased to call a Cube?

I declared that I could say nothing more, and that I must commit myself to the Truth, whose cause would surely prevail in the end.

The President replied that he quite concurred in my sentiment, and that I could not do better. I must be sentenced to perpetual imprisonment; but if the Truth intended that I should emerge from prison and evangelize the world, the Truth might be trusted to bring that result to pass. Meanwhile I should be subjected to no discomfort that was not necessary to preclude escape, and, unless I forfeited the privilege by misconduct, I should be occasionally permitted to see my brother who had preceded me to my prison.

Seven years have elapsed and I am still a prisoner, and—if I except the occasional visits of my brother—debarred from all companionship save that of my jailers. My brother is one of the best of Squares, just, sensible, cheerful, and not without fraternal affection; yet I confess that my weekly interviews, at least in one respect, cause me the bitterest pain. He was present when the Sphere manifested himself in the Council Chamber; he saw the Sphere's changing sections; he heard the explanation of the phenomena then given to the Circles. Since that time, scarcely a week has passed during seven whole years, without his hearing from

Episode. Hydesville was the location of a two-room farmhouse in New York state belonging to the Fox family. Strange raps at night had disturbed previous occupants, and "a shower of bumps and raps" occurred over several nights — always near the Foxes' daughters, eleven-year-old Katie (Katherine) and fourteen-year-old Maggie (Margaretta). Mrs. Fox had proved to her own satisfaction that the noises were made by a spirit: If she snapped her fingers once or did so twice, she elicited the same number of raps in reply. The unseen force could even answer questions: one rap for "no," two for "yes." Mrs. Fox attributed these phenomena to a spirit, and by 1850 she and her daughters were demonstrating their skills to the public as mediums. The practice of holding séances (meetings of small groups that revolved around a medium) to communicate with the spirits of the departed spread rapidly, and it became a popular pastime in Victorian England. Around 1890, Maggie Fox confessed that the whole thing had been a childish prank, a hoax perpetrated on her mother via an apple tied to a string. Later she withdrew the confession, but in any case it had had no effect at all on the spiritualist movement.

The main beliefs of the movement are that the dead pass into the "spirit world" and that a medium — a person sensitive to vibrations from the spirit world — can communicate with spirits. Spiritualists believe in a range of paranormal phenomena: telepathy (direct communication between minds), clairvoyance (the ability to see without using the eyes), clairaudience (the ability to hear without using the ears), materialization (the appearance of a spirit in our own material world), levitation (lifting an object by nonphysical means), apport (the nonphysical generation of objects), and teleportation (moving objects by the power of thought).

Spiritualism appealed to the Victorians for several reasons. Death was commonplace — the absence of contraception led to large families, but the rudimentary medicine of the period could not prevent or cure many common childhood or adult diseases. As the (sole) religious census of 1851 amply demonstrated, many Victorians were losing faith in traditional religion, in part because of its powerlessness in the face of death and disease, and were seeking a substitute that appeared to have more to offer. The rise of spiritualism was opposed by the established churches, by rationalists who argued that demonstrations of "psychic phenomena" were fraudulent, and by mobs who considered spiritualism a form of witchcraft.

The spiritualist movement attracted the favorable attention of Conan Doyle, author of the Sherlock Holmes and Professor Challenger stories, and Sir Oliver Joseph Lodge (1851–1940), who invented the "coherer," a detector for radio waves. Doyle turned to spiritualism after his son died from wounds during World War I; Lodge also lost a son in the trenches, but he was already a believer by that time. The events are chronicled in *Teller of Tales: The Life of Arthur Conan Doyle* by Daniel Stashower: Doyle was fooled by what to modern eyes were crude, unconvincing tricks. In 1917 he addressed the London Spiritualist Alliance, at a meeting chaired by Lodge, and said, "The subject of psychical research is one upon which I have thought more, and been slower to form my opinion about, than upon any other subject whatever....I have finally announced that I am satisfied of the evidence." This represented a major change of mind, because one year earlier, in the *International Psychic Gazette,* when asked for words of comfort for the families of soldiers killed in the war, he had said, "I fear I can say nothing worth saying. Time is the only healer." Doyle wrote several books on spiritualism, among them *The New Revelation* (1918) and *The Vital Message* (1919). James Douglas, writing in the *Sunday Express* newspaper, asked, "Is Conan Doyle mad?"

Matters came to a head in 1920 when Doyle wrote an article in *The Strand* magazine asserting the existence of fairies. Elsie Wright, a young girl from the Yorkshire village of Cottingley, had borrowed a camera from her father Arthur so that she

me a repetition of the part I played in that manifestation, together with ample descriptions of all the phenomena in Spaceland, and the arguments for the existence of Solid things derivable from Analogy. Yet—I take shame to be forced to confess it—my brother has not yet grasped the nature of the Third Dimension, and frankly avows his disbelief in the existence of a Sphere.

Hence I am absolutely destitute of converts, and, for aught that I can see, the millennial Revelation has been made to me for nothing. Prometheus up in Spaceland was bound for bringing down fire for mortals, but I—poor Flatland Prometheus—lie here in prison for bringing down nothing to my countrymen. Yet I exist in the hope that these memoirs, in some manner, I know not how, may find their way to the minds of humanity in Some Dimension, and may stir up a race of rebels[4] who shall refuse to be confined to limited Dimensionality.

That is the hope of my brighter moments. Alas, it is not always so. Heavily weighs on me at times the burdensome reflection that I cannot honestly say I am confident as to the exact shape of the once-seen, oft-regretted Cube; and in my nightly visions the mysterious precept, "Upward, not Northward," haunts me like a soul-devouring Sphinx. It is part of the martyrdom which I endure for the cause of the Truth that there are seasons of mental weakness, when Cubes and Spheres flit away into the background of scarce-possible existences; when the Land of Three Dimensions seems almost as visionary as the Land of One or None; nay, when even this hard wall that bars me from my freedom, these very tablets on which I am writing, and all the substantial realities of Flatland itself, appear no better than the offspring of a diseased imagination, or the baseless fabric of a dream.[5]

and her cousin Frances Griffiths could "take a picture of fairies." The children duly returned with a photograph showing Frances surrounded by diminutive people with wings. Her father, unimpressed, told her "you've been oop to summat," but two years later his wife showed the picture to her local branch of the Theosophical Society. (This society, founded in 1875 by Helena Petrovna Blavatsky [1831–1891] and Henry Steel Olcott [1832–1907], promoted a pseudoscientific brand of religious mysticism.) This brought the photo to Doyle's attention, and he became convinced that the image was genuine. Lodge was skeptical: In particular, the fairies' clothes and hairstyles reflected the latest Paris fashions. Despite these misgivings, in 1922 Doyle published *The Coming of the Fairies,* in which he wrote,

> It is hard for the mind to grasp what the ultimate results may be if we have actually proved the existence upon the surface of this planet of a population which may be as numerous as the human race, which pursues its own strange life in its own strange way, and which is only separated from ourselves by some difference of vibrations.

Accounts in the press were scathing and started using phrases like *easily duped.* Articles appeared pointing out that the Cottingley fairies looked remarkably similar to a picture on a popular brand of night light. Most of Doyle's spiritualist allies abandoned him because of the ridicule he was inflicting on their movement — guilt by association. Doyle never abandoned his belief that the photographs were genuine, although any modern viewer will spot them as fakes a mile off. It was all very sad, and it shows the extent to which rationality can be confounded by grief.

Doyle had long been a member of the Society for Psychical Research, founded in London in 1882 through the efforts of philosopher-psychologist William James (1842–1910). A number of prominent scientists also belonged. James was interested in religious experiences and in life after death, but he kept an open mind on such issues. A similar society started up in the United States in 1898. Today, a number of universities carry out research into "paranormal phenomena" under the designation parapsychology (memorably parodied in the film *Ghostbusters*). The subject remains controversial, and early research made major errors of methodology. One is worth a mention because it is mathematical. A typical experiment involved asking a subject to predict the symbols on a pack of special cards. Subjects who scored below average were removed from further consideration, whereas subjects who scored above average underwent further trials. Eventually a small proportion of subjects had racked up impressive scores, well above what would be expected if they were just guessing at random: These subjects were considered to have unusual paranormal talents. However, as experiments with these individuals continued, their performance slowly tailed off, back toward average … as though their powers were somehow waning.

In fact, these effects are exactly what should be expected if *all* subjects were guessing at random. On such a basis, some subjects will perform worse than average, and some will do better. Using probability theory, it is even possible to predict roughly how many, and how well or badly. The experimental procedure retains only those subjects who happen to have guessed better than average; the large numbers of discarded subjects are thereafter ignored. But now the experimenters are stuck with just those few people, and scorers below average are no longer discarded. From that moment on, the likely performance is merely average. But because the early successes, made during the selection procedure, are retained, they bias the subject's total score, and the effect of the later average scores sets in only gradually.

3 This passage is reminiscent of Galileo's appearance before the Inquisition, accused of heresy. According to the theory of the Alexandrian astronomer,

geographer, and mathematician Ptolemy (Claudius Ptolemæus, flourished 127–145), the sun, the moon, and all the planets orbited the Earth in a system of epicycles — circles whose centers revolved around other circles. The Roman Church favored the Ptolemaic cosmology because it placed the Earth at the center of the universe. However, in the period 1510–1514 a competitor emerged, when Nicolaus Copernicus (Niklas Koppernigk 1473–1543) circulated *Commentariolus* (*Minor Commentary*), a manuscript summary of a new theory in which the sun lies at the center of the solar system and the Earth and planets revolve around it. Aware of the dangers of inciting the wrath of the Church, Copernicus waited until 1540 before agreeing to publish his ideas. His *De Revolutionibus Orbium Coelestium* (*On the Revolutions of the Celestial Spheres*) appeared in 1543. It is said that a copy was presented to him on the day he died.

A 1597 letter to Kepler shows that Galileo was an early believer in Copernicus's heliocentric ("sun-centered") theory. In 1604 Galileo publicly acknowledged some aspects of the Copernican theory, and in 1613 he published three letters on sunspots, which he said proved Copernicus was right. The Church viewed these developments with growing dismay, and in 1615 the chief theologian Cardinal Robert Bellarmine (1542–1621, made a saint in 1930) warned Galileo to state his views more moderately. In 1616 Bellarmine persuaded the Congregation of the Index to suspend Copernicus's book. Galileo visited Rome twice to explain his position, and in 1624 Pope Urban VIII (1568–1644) gave him permission to issue hypothetical statements about "the system of the world." As a result, in 1632 Galileo published his *Dialogo Sopra i Due Massimi Sistemi del Mondo, Tolemaico i Copernicano* (*Dialogue Concerning the Two Chief World Systems: Ptolemaic and Copernican*), comparing the old, Earth-centered cosmology unfavorably with the new, heliocentric one. Galileo, who may have taken his permission to discuss hypotheticals a little too literally, was promptly summoned by the Inquisition. He observed that "In these and other positions ... it is not in the power of any creature to make them true or false otherwise than of their own nature ...," a sentiment that did not go down well with those in authority. He was probably not tortured physically, but the threat — even if only implicit — was enough: Pope Clement VIII (1536–1605) had condemned Giordano Bruno (1548–1600) to be burned at the stake for similar assertions. Galileo was forced to recant his scientific views and recite penitential psalms. He was sentenced to prison, but the sentence was immediately commuted to house imprisonment for the remainder of his life. In 1637, when he went blind, he was finally permitted visitors. Only in 1992 did the Church formally admit that it had been wrong to condemn him.

4 Abbott would surely be delighted by today's mathematics, in which space of four, five, a hundred, or indeed infinitely many dimensions is so commonplace as to be unremarkable to professionals, however bizarre it may seem to the uninitiated. He would also have applauded modern physics and the ten-dimensional universe postulated by the theory of superstrings. Did *Flatland* have any influence on these developments? It would be difficult to prove this either way, but it is notable that the City of London School produced a remarkable number of top-flight mathematicians and physicists (as well as classicists) during Abbott's tenure as headmaster, among them fifth wrangler L. B. Seeley, seventh wrangler G. F. Wright, senior wrangler W. S. Aldis, fourth wrangler F. Cuthbertson, sixth wrangler T. Skelton, third wrangler F. C. Wace, second wrangler F. Brown, eighth wrangler and first head of the Cavendish Laboratory John Cox, and second wrangler and First Smith's Prizeman J. E. A. Steggall. Many of them would surely have picked up a copy of their headmaster's witty and amusing book from the school library, and its contents may well have lodged in their minds, though perhaps subconsciously. (The copy

now owned by the school was presented to it on 16 June 1920 by R. H. Alpress, as is inscribed on the flyleaf, but in 1872 the departing master A. R. Vardy wrote, "Those who would like to taste Abbott in his lightest mood should certainly look for *Flat-land* [sic] in the School Library....") Who knows what consequences *Flatland* may have had for the development of mathematics? Certainly it remains popular among today's mathematicians.

5 To modern sensibilities, discovering that the events depicted in a story are merely a dream comes as something of an anticlimax (though, oddly, not for the previously mentioned example of *Alice's Adventures in Wonderland*). Perhaps Victorian readers found comfort in being told that the bizarre events of a story *never really happened,* not even in the fictional world being described, because such reassurance left their faith in the conventions of their own society intact. The cozy assumption that a doubly fictitious event is less real than a fictitious one scarcely bears close logical examination, but then neither does the modern reader's irritation at being told that manifestly fictitious events were a dream and thus "never really happened." It's fiction, we know that, right? The psychology of stories explains these apparently irrational attitudes. The reader of a story plays a psychological game: on the one hand, *believing that the story is real* in order to identify with its characters and gain the pleasure that goes with all good narratives; on the other hand, *knowing that it is not real* and thereby keeping in touch with reality. Followers of TV soaps who encounter actors in the street and berate or assault them because of the actions of the characters that they portray have failed to maintain this delicate balance. The distinction between SF fans and UFO buffs is similar: SF enthusiasts enjoy imaginative stories about aliens from outer space; ufologists are convinced the aliens are *here.*

In *Mind and the Universe,* Gregory Bateson (1904–1980) describes language as a hierarchy built up from simple structures, such as nouns and verbs, to more complex ones, such as paragraphs, chapters, and novels. From this point of view, the distinction between fact and fiction lies at the highest intellectual level. If a member of the cast is shot on stage during the action of a play, nobody in the audience phones the police. (But then there is the man who jumped up and shouted, "He's lying, you idiot!" during a performance of *Othello*.) Bateson waxes lyrical about this "highest function" of language. Then he tells his readers that, feeling really satisfied with having written those pages, he rewarded himself with a visit to the zoo. Immediately inside the front gate he came upon two monkeys in a cage, *playing* at fighting ... and everything turned end over end. The monkeys had no *word* for fiction, but they knew what "pretend" meant. Bateson then recasts his theories, making the distinction between fact and fiction the *basis* of language, not its pinnacle.

6 In Shakespeare's *The Tempest* (act 4, scene 1, lines 148–158), Prospero says,

> These our actors,
> As I foretold you, were all spirits, and

Are melted into air, into thin air;
And like the baseless fabric of this vision,
The cloud-capp'd towers, the gorgeous palaces,
The solemn temples, the great globe itself,
Yea, all which it inherit, shall dissolve,
And, like this insubstantial pageant faded,
Leave not a rack behind. We are such stuff
As dreams are made on; and our little life
Is rounded with a sleep.

7 See above.

8 See above; the precise quote is "... as dreams are made *on*."

9 The figure in the first edition is tilted relative to this version; in the first edition, the phrases THE END and FLATLAND run horizontally. At the foot of the figure the first edition has a horizontal line and underneath the legend LONDON: R. CLAY, SONS, AND TAYLOR, PRINTERS. This firm, of course, was the printer of the first edition. It also printed the second and many issues of the *City of London School Magazine*. Its address was Bread Street Hill, London.

THE FOURTH DIMENSION
IN MATHEMATICS

The mathematical path to the world of four dimensions — and beyond — was surprisingly indirect and tortuous. Today, we define four-dimensional space as the set of all quadruples (x_1, x_2, x_3, x_4) of numbers, by analogy with the coordinate representation of spaces of two and three dimensions. Then we define the distance between points using an equally straightforward analogue of the Pythagorean theorem, and everything else follows from that. To modern mathematicians this approach seems almost obvious — and equally obviously, the resulting concept is a space just as real as the mathematician's three-dimensional space … which is to say, not real at all. Why did it take so many centuries to reach such a simple description? Perhaps in earlier times today's approach to four dimensions was *too* obvious, too simple to look interesting or useful to anyone who happened to think of it, and too divorced from reality even if they did. The important simplicities of mathematics are often extracted from more complicated things, by gradually stripping away irrelevant detail. And often, new concepts arrive by a circuitous route: Investigations initially carried out in one area of mathematics turn

out to have major implications elsewhere. The key to the fourth dimension came from algebra, not geometry, and the original intention was to find an algebra of three dimensions, not four. But that algebra built on a solid foundation of geometric thinking, and it led to a new vision of what geometry should be, so the story must begin and end with geometry.

The Origins of Geometry

The word *geometry* comes from the Greek words γη (*ge,* "Earth") and μετρια (*metria,* "measuring"), hence "land measurement." For most practical purposes, land is two-dimensional; for example, when you buy a house, the lawyers are happy to work with a two-dimensional plan. To the early Greeks, *geometry* referred only to the plane, a space of two dimensions. Later they added two other spaces: the "solid geometry" of three-dimensional space, and the "spherical geometry" of triangles and other shapes on the surface of a sphere. In all three cases, though, geometry was about the space that we inhabit — possibly, in the case of the plane and sphere, some lower-dimensional and therefore simpler part of that space. In Greek geometry, it was assumed that physical space is of the kind that we now call Euclidean, after the great Greek geometer Euclid, whose 13-volume work *The Elements* collected, formalized, and systematized a substantial portion of the geometry known in his day. Recall that two straight lines in the plane are parallel if they do not meet no matter how far they are extended — an infinite version of the two rails of a railway track. A key feature is that in Euclidean spaces, parallel lines exist, and the Greeks considered this to be such an obvious feature of our physical universe that they accepted it without demur.

The sphere, of course, is not Euclidean in this sense. In fact, there are no straight lines on a sphere, but the analogous concept is a great circle, a circle that lies in a plane through the center of the sphere (Figure 43). Parallel lines do not meet, but *any* two great circles meet. Therefore, there are no parallels on a sphere. However, a sphere is visibly not a plane, so the Greeks were not greatly bothered by this fact.

Spherical geometry was, and still is, important for astronomy and navigation, because the positions of the stars and planets lie on a conceptual sphere — the celestial sphere — and the Earth is within 0.3% of a perfect sphere. Euclid's *Phaenomena* (c. 300 B.C.), an astronomy text, includes some spherical geometry. Around 20 B.C., Theodosius assembled much of what was known about spherical geometry in his *Sphaericae* (*On Spheres*). A later important contributor was Ptolemy (Claudius Ptolemaeus), who employed spherical geometry as a

Figure 43 Great circles on the sphere always meet.

basis for astronomy in his *Syntaxis Mathematica* (*Mathematical Collection*), a book that the Arabs referred to as the *Almagest* (*The Great* [*Collection*]). Ptolemy's work is theoretical: Its main contact with practical issues is an argument *against* the heliocentric theory — the theory that the Earth revolves around the sun — which had been suggested by Aristarchus (c. 310–230 B.C.). Ptolemy did such a good job that little changed in spherical geometry for the next thousand years. His ideas were preserved and developed mainly in the Arab world, in particular by Abû'l-Wefâ (939–997) and Nasîr-Eddin (1201–1274).

The key to spherical geometry, as for Euclidean geometry, is a good understanding of triangles — spherical *trigonometry*. The essentials of the subject returned to Europe around 1462, when Regiomontanus (Johannes Müller, 1436–1476) wrote *De Triangulis* (*On Triangles*). Based on the writings of Nasîr-Eddin, it was published in 1533. Meanwhile, it was improved by Johann Werner (1468–1528) in his *De Triangulis Sphaericis* (*On Spherical Triangles*) of 1514. The big problem with spherical trigonometry is a rather large collection of basic formulas that must be committed to memory:

Some major simplifications were introduced by François Vieta (1540–1603), starting with his *Canon Magnus* (*Great Rule*) of 1579. By the end of the sixteenth century, spherical geometry was fast becoming an area of mathematics in its own right, rather than just a sub-branch of astronomy.

In each of these three incarnations, "geometry" was directly related to physical space or, more accurately, to a presumed model of physical space. The same went (or so at first it seemed) for the next development, projective geometry. Projective geometry is closely connected with techniques for obtaining realistic perspective in painting. Early ideas of artists such as Filippo Brunelleschi (1377–1446) and Paolo Ucello (1397–1475) were developed into a precise mathematical system by Leone Battista Alberti (1404–1472) in his 1435–1436 book *Della Pittura* (*On Painting*), first printed in 1540. Alberti pointed out that a picture on a canvas is a section of a projection of the real scene. That is, imagine every point in the scene to be joined by a line to the artist's eye. Now slice this bundle of lines with a flat plane (the canvas), and each line meets the plane in a point that corresponds to the image of the original point in the scene. The essence of projective geometry is the following question: Which features of a geometric figure are preserved by projection and section? As posed, this question is still related to ordinary physical space, but in modern times it became clear that the underlying space for projective geometry has its own characteristic structure. Essentially, it is the geometry of the surface of a sphere in which an-

tipodal (diametrically opposite) points are considered to be identical. If these point-pairs are replaced by literal points, no such space exists in three-dimensional Euclidean space: It is a conceptual construct.

Non-Euclidean Geometry

Four-dimensional geometry could not be invented while the accepted intuition was that geometry is directly related to physical space, and the overthrow of this point of view took a long time, probably because it is such a natural one. Geometry was freed from any *necessary* connection to physical space only in the nineteenth century, although there are some intriguing earlier foreshadowings of this development. The story again goes back to Euclid, but this time what matters is the *logical* structure of the *Elements*.

Euclid's most far-reaching contribution to mathematics was to introduce a method that remains the basis of all mathematical concepts to this day, the *axiomatic* method. Each concept is defined by stating a series of properties — axioms — that it will be assumed to possess. All subsequent developments are logical consequences of these axioms, and a "proof" is an explicit argument that shows how these consequences arise and why they are logically valid. The resulting theory is a logically ordered sequence of properties, all of which are implicit in the axioms; the mathematician's job is to make them *explicit*. Anything that satisfies the axioms must necessarily possess all the properties in the list as well. The advantage of the axiomatic method is that it leaves mathematicians free to find out how various concepts *behave*, without worrying about tricky philosophical issues of what they *are*. Instead of asking what things satisfy the axioms, mathematicians focus on what else is true of anything that does. Proof thus becomes the core requirement of mathematics, just as agreement with experiment is the core requirement of the theoretical sciences.

Euclid's approach is elegant and economical: He founded the entire elaborate edifice of Greek geometry on a small number of axioms (to modern eyes, too small). Most of these were statements so "obvious" that they could scarcely be challenged, such as "Any two points can be joined by a straight line." But one, later named the Parallel Axiom, seemed far less transparent: "If a straight line falling on two straight lines makes the interior angles on the same side less than two right angles, the two straight lines, if produced indefinitely, meet on that side where the angles are less than two right angles" (see Figure 3). There were two main approaches to the removal of this perceived defect: to replace the Parallel Axiom by something more intuitive, or (better, if possible) to deduce it from the other axioms and so render it superfluous.

The first major attempt was made by Ptolemy in a (lost) tract on the Parallel Axiom, but his alleged proof foundered on a subtle unproved assumption. Nevertheless, in the fifth century, the commentator Proclus (410–185) stated that the Parallel Axiom was really a theorem proved by Ptolemy (the flaw was not then readily apparent) and offered his own

proof — but this proof was based on the unproved assumption that two lines that cross each other will become farther and farther apart as they get farther away from the crossing point. In fact, this assumption is logically equivalent to the Parallel Axiom: It is more intuitive, perhaps, but Proclus had not deduced it correctly from the other axioms, as he imagined. Nasîr-Eddin, who edited Euclid for Persian readers, offered his own proof, but it was based on essentially the same assumption. John Wallis (1616–1703) proved the Parallel Axiom by assuming the existence of similar triangles, triangles with the same angles but different sizes. Again, this may be a more plausible axiom, but it did not occur to Wallis that it, too, needed proof. In 1769 Joseph Fenn suggested what is probably the simplest logical equivalent to Euclid's Parallel Axiom: "Two intersecting lines cannot both be parallel to a third line." This in turns is easily seen to be equivalent to the best-known substitute, Playfair's axiom: "Given a line, and any point not on that line, there exists one and only one line parallel to the given line and passing through the given point."

At about this time in history, the mathematical heavyweights began to get interested in the question. Adrien-Marie Legendre (1752–1833) included a discussion in his *Eléments de Géometrie* (*Elements of Geometry*). Restricting himself to Euclid's axioms minus the Parallel Axiom, he gave a correct deduction of what subsequently turned out to be a key theorem: If some triangle has an angle sum that is equal to 180°, then the same is true of every triangle; and if some triangle has an angle sum that is less than 180°, then the same is true of every triangle. (It follows at once that if some triangle has an angle sum that is greater than 180°, then the same is true of every triangle.) Finally, if some triangle has an angle sum of 180°, then the Parallel Axiom is true. In short, there are three distinct cases to consider, precisely one of which implies the Parallel Axiom. If the other two cases could be eliminated on the grounds of logical inconsistency, then the Parallel Axiom would become a theorem and could be removed from Euclid's list of axioms.

Soon thereafter, the Jesuit priest Girolamo Saccheri (1667–1733) published *Euclides ab Omni Naevo Vindicatus* (*Euclid Freed of Every Flaw*), which attempted to do just that. He based his approach on a variant of Legendre's observation. Construct (see Figure 44) a quadrilateral ABCD in which A and B are right angles and AC = BD. Then, without using the Parallel Axiom, it can be proved that angles C and D are equal. Saccheri pointed out that there are three possibilities:

- C is a right angle.
- C is greater than a right angle.
- C is less than a right angle.

The first case implies the Parallel Axiom. Saccheri investigated the second case (the Hypothesis of the Obtuse Angle), making a long series of correct deductions, the last one of which seemed to him so counterintuitive that he rejected it. He did much the same with the third case (the Hypothesis of the Acute Angle), which (he assumed) enabled him to claim that his title was justified. Nonetheless, he

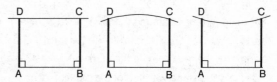

Figure 44 Saccheri's three cases.

had not actually disproved either the second or the third case on the basis of the other axioms of Euclid, and this became recognized by the experts. By 1759 d'Alembert said that the problem of the Parallel Axiom was "the scandal of the elements of geometry."

Some, however, had seen faint glimmerings of light. In 1763 Georg Klügel (1739–1812), who had read Saccheri's book, made the prescient comment that the intuitive plausibility of the Parallel Axiom was based on experience, not on logic. Specifically, Klügel said that Saccheri had not disproved his Hypotheses of the Obtuse Angle or the Hypothesis of the Acute Angle; he had merely made deductions that seemed contrary to experience. For the first time, it was beginning to dawn on mathematicians that perhaps the Parallel Axiom could not be proved from the others. In his 1766 *Theorie der Parallellinien* (*Theory of Parallel Lines*), Johann Heinrich Lambert (1728–1777) followed Saccheri in rejecting the Hypothesis of the Obtuse Angle, but he suggested — on the basis of a formal analogy with spherical trigonometry — that the Hypothesis of the Acute Angle would be valid for the geometry of a sphere of radius $\sqrt{-1}$. Because no such sphere exists, however, this suggestion remained obscure and did not really solve the problem. Its obscurity notwithstanding, the idea was taken up by others, among them Ferdinad Schwei-kart (1780–1859), who introduced the term astral geometry, the geometry of the stars.

Finally the stage was set for the creation of non-Euclidean geometry, which came about independently through the work of three men: Carl Friedrich Gauss (1777–1855), János Bolyai (1802–1860), and Nikolai Ivanovich Lobachevskii (1793–1856). Gauss worked out his ideas on the Parallel Axiom between 1792 and 1816 but never published anything. Bolyai's work stems from about 1825 and was published as an appendix, "The Science of Absolute Space," to his father Wolfgang Farkas Bolyai's 1832 book *Tentamen Juventutem Studiosam in Elementa Matheosos* (*Essay on the Elements of Mathematics for Studious Youths*). Lobachevskii published his work in 1829. All three mathematicians had realized — though none of them had a complete proof — that there are geometries in which Euclid's Parallel Axiom fails but all the other axioms are valid. The work of Saccheri, for example, largely consisted of correct proofs of basic theorems in these *non-Euclidean* geometries. Indeed, there were two types of non-Euclidean geometry:

- Hyperbolic geometry, which obeys Saccheri's Hypothesis of the Acute Angle
- Elliptic geometry, which obeys Saccheri's Hypothesis of the Obtuse Angle

Gauss was something of a physicist as well as a mathematician; he was convinced that the geometry of real space was Euclidean and wanted to prove it experimentally. He therefore measured the angles between the three mountain peaks of Brocken, Hohehagen, and Inselsberg — a triangle with sides of 69km, 85km, and 197km. His measurements showed an angle sum of 180° 0' 14.85", but the discrepancy of 14.85" was well below the likely level of experimental error, so the result could not distinguish among Euclidean, hyperbolic, and elliptic geometry. In fact, Gauss's method could be conclusive only if space is *not* Euclidean.

Somewhat later, it was shown that the two types of non-Euclidean geometry could be interpreted as the geometry of geodesics (shortest paths) on surfaces of constant positive or negative curvature — positive for elliptic geometry, negative for hyperbolic geometry. The Euclidean plane has zero curvature, a sphere has positive curvature, and a saddle-shaped surface has negative curvature (Figure 45). Gauss had studied the curvature of surfaces, and he had devised a beautiful formula to calculate its value. He took great pride in this formula, calling it his *theorema egregium,* Most Excellent Theorem. The formula is based solely on the intrinsic geometry of the surface; that is, it is independent of any specific way in which the surface might be embedded in three-dimensional space. For example, if a plane, which has zero curvature in Gauss's sense, is rolled up into a cylinder, then the curvature remains zero.

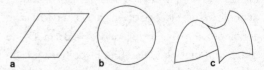

Figure 45 Surfaces with (**a**) zero curvature, (**b**) positive curvature, and (**c**) negative curvature.

An important implication is that hyperbolic and elliptic geometry can also be given an intrinsic meaning without invoking Euclidean geometry. It is then possible to prove that although the Parallel Axiom fails in these geometries, all of the other Euclidean axioms are valid. It took a while for this realization to dawn, mainly because the question had to be stated unambiguously, but when it did, it became clear that the whole issue could in principle have been resolved by the Greeks. It takes a page or so to explain what is involved, but the story is highly illuminating. The main conceptual leap is to introduce new terminology into spherical geometry:

- LINE = great circle
- POINT = pair of antipodal points

Next, it is necessary to establish that nearly all of Euclid's axioms are valid when *line* is replaced by LINE and *point* by POINT. This requires little more than routine verifications, couched in the new "language." For example, an old-fashioned triangle becomes a TRIANGLE, which is a pair of mutually antipodal spherical triangles. None of this is hard, but it is uncomfortable if you confuse the new language with the old.

Why go through these strange antics? Suppose someone claims to have a proof of the Parallel Axiom based solely on the other axioms. It is, perhaps, a complicated proof — say, 2001 statements long. Each statement is alleged to be a logical consequence of the previous ones, together with all of the axioms save the Parallel Axiom. Working through the proof step by step, looking for a flaw,

would be very hard work, and a subtle flaw might easily be missed. Thanks to the reinterpretation of spherical geometry in terms of LINEs and POINTs, however, that's not necessary. You can prove that there must be a flaw, a logical error, without locating it — indeed, without reading the proposed proof. This kind of thing enrages the proud owner of the proof, whose careful work is totally ignored yet declared erroneous; however, it is not arrogance or authoritarianism — just simple logic. Here's how it works.

The proof, whatever it is, talks of lines and points (in the Euclidean sense) and concludes with the existence of parallel lines — lines that never meet. So far, so good. But now we translate the proof word for word into a proof in spherical geometry, replacing every occurrence of *line* by LINE and every occurrence of *point* by POINT. The proof remains valid in this new interpretation because all of Euclid's axioms except the Parallel Axiom are satisfied, and the essence of what we are saying about the proof is that it never uses the Parallel Axiom. So what theorem in spherical geometry does it prove? That there exist parallel LINEs — ones that do not meet. However, it is a standard theorem of spherical geometry, which can also be proved without using the Parallel Axiom, that any two great circles (LINEs) always intersect. These two statements contradict each other, yet each is (allegedly) derived from the same axioms: Euclid's axioms minus the Parallel Axiom.

In summary, *either* spherical geometry is logically inconsistent in its own right, *or* the purported

2001-step proof of the Parallel Axiom contains a mistake.

As a final twist, spherical geometry is really just a special part of three-dimensional Euclidean geometry, which in turn is an elaboration on two-dimensional geometry. Thus there are two possibilities: Euclidean geometry is nonsense, or the proof of the Parallel Axiom is wrong. Neither possibility "frees Euclid of every flaw." Indeed, if the alleged proof of the Parallel Axiom is correct, then Euclidean geometry is destroyed.

I repeat that it doesn't *matter* how the 2001-step proof goes. The foregoing argument works whatever the proof may be. It rests solely on what the proof is claimed to have achieved. The conceptual point behind this argument is that spherical geometry provides a "model" for Euclid's axioms minus the Parallel Axiom — that is, an interpretation in which those axioms are valid. It follows that any correct logical deduction from those axioms must also be valid. Because the Parallel Axiom is not valid in the model, it can't be a correct logical deduction.

The most important conclusion of this argument — along with a similar one for hyperbolic geometry, which uses a less familiar but still straightforward model — is that the Parallel Axiom is logically independent of the other axioms, and cannot be deduced from them. Euclid's intuition that such an axiom was needed is vindicated. Moreover, geometry in the mathematical sense now becomes divorced from the nature of physical space. Mathematically, a geometry makes sense provided that it is logically coherent and does not lead to contradictions. The physical nature of space is a matter for experiment. Our senses help us to formulate possible mathematical models for physical space and to devise appropriate experiments, but our senses may also be misleading, because our mental image of space is based on limited experience in one tiny region of a vast universe.

The Complex Plane

The discovery of non-Euclidean geometry drove home the message that mathematicians are free to invent new kinds of geometry, without these being required to represent some perceived model of physical space. The new geometric spaces were still two-dimensional, however; there *are* three- (or higher-) dimensional generalizations, but those came rather later. Mathematics, then, had not yet escaped the straitjacket of Euclidean dimensionality, but it had escaped the Prison of Parallels.

In the meantime, a revolution had also been brewing in the field of algebra, which eventually sparked the creation of geometries of four (or more) dimensions. Geometry as conceived by the Greeks was essentially a formalization of the human sense of sight and touch, which between them allow our brains to build internal models of the positional relationships of the outside world. Geometry took its first tentative steps toward freeing itself from this limited interpretation when the algebraists of Renaissance Italy unwittingly stumbled on a new concept of number.

Algebra is — or, at least, began as — the art of solving equations. Given an unknown quantity, or several, and some conditions that must be satisfied by that quantity, algebra provides a battery of methods for discovering the value of the unknown quantity or quantities. In fact, the name comes from the Arabic phrase *al-jabr,* an abbreviation of *Al-jabr w'al-muqâbala.* This phrase, which is the title of a text by Mohammed ibn Musa al-Khôwarizmî (c. 825), means "the art of restoration and simplification" and hence refers to two of those methods. Al-Khôwarizmî is also immortalized in the word *algorithm* (used in computer science to mean a specific computational procedure or "program"), which is derived from his name.

Algebra began as a series of verbal rules for finding "the unknown." These rules were often described by examples. For instance, suppose that we are told

- When the square of the unknown is added to three times itself and seven is subtracted, the result is nine.

Then the verbal rule to find the unknown would be to modify that statement by carrying out the following steps:

- *(Restoration)* When the square of the unknown is added to three times itself, the result is sixteen.
- *(Simplification)* Four times the square of the unknown is equal to sixteen.
- *(Simplification again)* The square of the unknown is equal to four.

It is then obvious that the unknown is equal to two, and the problem is solved.

Other rules were less obvious. For example, Al-Khôwarizmî discusses the following problem:

A square and ten of its roots are equal to nine and thirty dirhems; that is, when you add ten roots to one square, the sum is equal to nine and thirty.

His solution reads,

Take half the number of roots, that is, in this case five; then multiply this by itself and the result is five and twenty. Add this to the nine and thirty, which gives sixty-four; take the square root, or eight, and subtract from it half the number of roots, namely five, and there remains three. This is the answer.

It is easy to see that the *answer* is correct, but why is the *rule* correct? For that, it is simplest to employ modern notation. Algebra really began to take off with the introduction of symbolic representations, but it was a long time before these evolved into today's streamlined symbolism. What we now do is use letters of the alphabet to represent unknowns, the standard choice for a single unknown being *x.* The square of *x* is then written x^2, and the first example above then becomes the equation

$$x^2 + 3x^2 - 7 = 9$$

This is solved by changing it into

$$x^2 + 3x^2 = 16$$

(add 7 to both sides), then into

$$4x^2 = 16$$

(collect the terms in x^2), and finally into

$$x^2 = 4$$

(divide both sides by 4). This leads to the solution $x = 2$.

The second example is more complicated. The problem translates into

$$x^2 + 10x = 39$$

The first step, which comes like a rabbit out of the conjuror's hat, is to add $5^2 = 25$ to each side:

$$x^2 + 10x + 25 = 64$$

Today we recognize the left-hand side of this as the square of $x + 5$ (which is where the rabbit comes from), and we can rewrite the equation as

$$(x + 5)^2 = 64$$

so that

$$x + 5 = 8$$

and $x = 3$. Al-Khôwarizmî's verbal description is an exact parallel. The technique involved, "completing the square," remains in use to this day. However, nowadays we recognize that 64 has *two* square roots, 8 and -8, so there is a second solution: $x = -13$. Indeed, every positive number has two square roots, zero has just one, and negative numbers do not have square roots because the square of any number is positive. At least, this is the case for the familiar numbers of school arithmetic, and thereby hangs a tale.

An equation such as the above, involving the first and second powers x and x^2 of the unknown, is called a *quadratic* equation. The key to the fourth dimension — as things turned out on this planet, at least — is the next simplest kind of algebraic equation, which is called a *cubic* equation. In addition to the first and second powers, a cubic equation also involves the cube x^3 of the unknown, and it's *much* harder to solve. Al-Khôwarizmî's technique of com-

pleting the square was essentially known to the Babylonians in 2000 B.C. Methods for solving cubic equations had to wait until the Renaissance and the work of Scipione del (or dal) Ferro (or Ferreo) (1465–1526) and Niccolò Fontana (1499?–1557). Del Ferro solved a significant class of cubic equations around 1500 but did not publish his method. In those days, mathematicians often engaged in public contests to demonstrate their prowess, so publication of anything radically new was not an option. However, around 1510 del Ferro passed his method on to Antonio Maria Fior (c. 1500– c. 1550) and to Ferro's son-in-law Annibale della Nave (1500?–1558).

Enter Fontana, nicknamed Tartaglia, "the stammerer," for a condition brought about by a slash across the face from a French soldier's saber when he was a boy. Tartaglia discovered how to solve all cubic equations, not just del Ferro's special class, and in 1535 he wiped the floor with Fior in a public competition by setting his rival only problems in the class that he could not solve. Tartaglia unwisely divulged his methods to Gerolamo Cardano (1501– 1576), who in addition to being a formidable scholar, was also a rogue and a gambler. Cardano promptly gave away the secret in his *Ars Magna* (*The Great Art*) of 1545 — with full credit to its inventor, but that just added insult to injury. This classic algebra text also explained a method that Cardano's pupil Lodovico Ferrari (1522–1565) had discovered for solving *quartic* equations, those that involve the fourth power x^4 of the unknown.

It was Cardano who set mathematics on a path

that led to an innovation of astonishing significance, with the innocent-looking problem

$$x + y = 10$$
$$xy = 40$$

in two unknowns x and y. In the *Ars Magna* he manipulates these equations (in the notation of the time) to obtain solutions that we would now write as

$$x = 5 + \sqrt{-15}$$
$$y = 5 + \sqrt{-15}$$

Traditionally, -15 cannot have a square root (so the equations have no solution, a not uncommon phenomenon), and Cardano offered no explanation of what the square root of a negative number might mean. He did observe that *if* such a number obeys the usual algebraic rules, then x and y are indeed solutions, but he seemed unimpressed by the fact, saying, "So progresses arithmetic subtlety, the end of which is as refined as it is useless." However, Cardano also ran into square roots of negative numbers in circumstances where they could not be so easily dismissed: the cubic. He noticed that when Tartaglia's method is applied to the cubic equation

$$x^3 = 15x + 4$$

it leads to the apparently nonsensical result

$$x = \sqrt[3]{2 + \sqrt{-121}} + \sqrt[3]{2 - \sqrt{-121}}$$

However, there is an easy answer, $x = 4$, which Tartaglia's method does not find.

Raphael Bombelli (1526–1573) noticed that maybe Tartaglia's method does work here after all, but for a very curious reason. Assuming that the square roots of negative numbers obey the usual algebraic rules, a short calculation shows that

$$(2 + \sqrt{-1})^3 = 2 + \sqrt{-121}, \text{ whence}$$
$$\sqrt[3]{2 + \sqrt{-121}} = 2 + \sqrt{-1}$$
$$(2 - \sqrt{-1})^3 = 2 - \sqrt{-121}, \text{ whence}$$
$$\sqrt[3]{2 - \sqrt{-121}} = 2 - \sqrt{-1}$$

so that Cardano's expression, obtained from Tartaglia's formula, can be rewritten as

$$x = 2 + \sqrt{-1} + 2 - \sqrt{-1} = 4$$

thereby recovering the known solution by a remarkably circuitous and puzzling route. Bombelli was thus the first person to establish the possibility that square roots of negative numbers might be useful in mainstream mathematics, if only as an intermediary toward something else. Neither Bombelli nor any of his contemporaries could explain *why* these calculations gave a correct answer, though, and the idea languished. But it did not go away: Like a seed in the desert waiting for a thunderstorm, it lay dormant until conditions were right for its germination.

The first of these conditions was the invention of calculus, achieved more or less independently by Newton and Leibniz. The central idea of the calculus is to study the instantaneous rate of change of some quantity that varies with time. For example the rate of change of the position of a body is its velocity, and the rate of change of velocity is acceleration. The two key operations of calculus are differentiation (given a quantity, find its rate of change) and integration (given the rate of change, find the quantity). Leibniz's main motivation was mathematical and philosophical, whereas Newton was hot on the trail of applications to physics: his Laws of Motion and his Law of Gravitation.

The second condition developed out of Newton's work: the realization that nature can be described by mathematical equations, leading to a growing need for good techniques for *solving* those equations. The equations of mathematical physics extended far beyond motion and gravity; they apply to heat, light, sound, and fluid mechanics.

The third condition was a growing awareness of the importance of logical rigor — a rediscovery of the Greek emphasis on *proof*. Mathematics is not an experimental science, and it is not a good idea to accept a mathematical assertion merely because it is true in a large number of examples. No mathematical concept can be considered established until it has been given a precise logical definition and its main properties have been proved on the basis of that definition.

Around 1700 several prominent mathematicians were exploring calculus, mainly from the point of view of mathematical physics, and they discovered intriguing connections with Bombelli's bizarre algebra of nonexistent square roots. John Bernoulli (1667–1748), Leibniz, and Euler found elegant uses for logarithms of such numbers, although there was some controversy about which results were correct. A remarkable consequence was Euler's famous formula

$$e^{\pi\sqrt{-1}} = -1$$

relating $\pi = 3.14159...$ and $e = 2.71828...$ (the "base of natural logarithms' and a number as fundamental to analysis as π is to geometry). Other wonderful and useful formulas could be derived from this, and many of them could be checked by other means. The operations of calculus could be extended into this new realm. The results were puzzling and at times seemed self-contradictory, but the consequences included powerful new techniques for solving the equations of mathematical physics. Moreover, the answers could be checked independently of their method of derivation — and they were *right*. The mysterious new method seemed to give good answers, even though nobody was quite sure why it worked or what its logical basis might be. The need for a logical theory of the square root of -1 was becoming pressing.

The route to this new number system began with a single leap of faith, after which everything else could be deduced by following one's mathematical nose in a fairly straightforward manner. The starting point was the square root of -1. The key to progress was to stop worrying about what this baffling statement *meant* and just see what could be done with it if it *did* have a meaning. A useful step was to introduce a new symbol to represent this "number." The historical choices varied, but around 1776 Euler introduced the symbol

$$i = \sqrt{-1}$$

and Gauss made it standard. Today's engineers prefer j to avoid conflicts with established uses of i in their subject, but mathematicians stick to Euler's choice. By definition, i satisfies the equation

$$i^2 = -1$$

and that's pretty much all we know about it. As a reminder that i isn't a normal "real" number, call it *imaginary*. We want to do algebra with it, and in particular we will need to combine it with the usual "real" numbers, so along with *i* there must be other baggage, like 3*i* and 5 + 3*i*. Numbers of the form

$$x + iy$$

with *x* and *y* real are said to be *complex* (in the sense "composed of several parts" rather than "complicated"). That step having been taken, it turns out that there is no need to go further. If you add two complex numbers together, you get a complex number. The same goes if you multiply them or even divide one by another. *Every* complex number other than 0 has exactly two square roots, one being minus the other. Indeed, every solution of a polynomial equation with complex coefficients is a complex number. In short, once you've thrown in *i*, you definitely need all the complex numbers too, but whatever operations of arithmetic and algebra you then wish to employ, it is not necessary to enlarge the system of numbers even further. It is as though the complex numbers have been waiting all along as a necessary extension of the traditional real number system — and they do the job so effectively that they are the *only* extension you need.

Fine — but what manner of beast is *i*?

One way to answer that question is to declare *i* meaningless and move on to other areas of enquiry. Some mathematicians and philosophers did just that, but others had a feeling that to do so was to miss out on a creative new idea, and evidence gradually accumulated in favor of that view. The realization that these strange new kinds of number were too useful to allow them to be dismissed as mere accidents or fantasies took several centuries, but slowly they entered the common mathematical

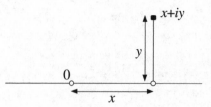

Figure 46 Complex numbers as points in the plane.

consciousness. In his 1637 *La Géometrie* (*Geometry*), René Descartes (1596–1650) distinguished ordinary "real" numbers from these newfangled "imaginary" ones, but he went only as far as suggesting that the presence of imaginaries implies the absence of solutions (a viewpoint that Bombelli's work refutes). In his *Algebra* of 1673, John Wallis (1616–1703) devised a geometric representation of complex numbers *x* + *iy*. Fix a line in the plane, and measure off the "real part" *x* along that line; then measure off the "imaginary part" *y* in a direction *at right angles to the line* (Figure 46). In this manner, every complex number corresponds to a unique point of the plane.

This suggestion seems to have been totally ignored, although its later rediscovery (see below) is considered to have been highly significant. Presumably, Wallis was just too far ahead of his time. Three mathematicians independently rediscovered a slightly improved version of Wallis's geometric representation: that the complex number *x* + *iy* corresponds to the point in the plane with Cartesian coordinates (*x, y*). The first was Caspar Wessel (1745–1818), a Dane, who published in 1797. His work was, naturally, in Danish, so it went unnoticed until a French translation appeared a century later. Independently, the French mathematician Jean-Robert Argand (1768–1822) published the same idea in 1806, and the geometric representation of complex numbers is often called the *Argand diagram*. Around 1811 Gauss had also come to view the complex numbers as points in a plane, and he wrote a letter to that effect to his friend Friedrich Wilhelm Bessel (1784–1846). Gauss published the idea in 1832, lending an air of respectability to the whole enterprise.

Finally, in 1837 — nearly 300 years after Cardano's remarks about "refined and useless" subtlety — William Rowan Hamilton (1805–1865) reduced the whole topic to simple algebra. *Define* a complex number $x + iy$ to be a *pair* of real numbers (x, y). Define addition and multiplication of these pairs by the rules

$$(x, y) + (u, v) = (x + y, u + v)$$
$$(x, y)(u, v) = (xu - yv, xv + yu)$$

That's it. It is possible to check, one rule at a time, that these pairs obey all the standard algebraic rules. Moreover, pairs of the form $(x, 0)$ mimic the arithmetic of the real numbers and so can be replaced by the symbol x. Euler's number i can be defined to be

$$i = (0, 1)$$

because

$$(0, 1)(0, 1) = (-1, 0)$$

which is the same as plain -1. Finally,

$$x + iy = (x, 0) + (0, 1)(y, 0) = (x, y)$$

so the new (x, y) is consistent with the old system of symbols. And if you *want* to, you can represent the pair (x, y) as a point in the plane, but this interpretation is not actually necessary to establish a meaning for complex numbers. They are just pairs of real numbers, equipped with algebraic rules that are motivated by the need to provide a square root for the pair $(-1, 0)$. Easy!

Four-Dimensional Numbers

Complex numbers equip the homely plane with an algebraic structure, and complex analysis — the analogue of calculus — makes it possible to solve many problems related to the mathematical physics of systems in the plane. However, we live in three-dimensional space. Complex numbers were such a good trick for any kind of problem in the plane that a similar trick for three-dimensional space would be invaluable. Accordingly, mathematicians expended a lot of effort trying to find a *three*-dimensional number system, which they tacitly assumed would have to satisfy all the usual laws of algebra, in the hope that the associated calculus would solve important problems of mathematical physics in three-dimensional space. Try as they might, they could not find such a number system. Eventually, someone discovered why.

Among the "usual laws of algebra" is the *commutative law of multiplication*, which states that $ab = ba$. It is to Hamilton that we owe a daring insight: This law can, if necessary, be abandoned. In fact, if you want an algebraic formalism that corresponds to the geometry of more than two dimensions, it *has* to be abandoned. This is much clearer now, with the benefit of hindsight: The key issue is *rotations*. In the plane, if you rotate an object through some angle *a* about the origin, and then through some angle *b*, the combined effect is a rotation through angle *a* + *b*. If the rotations are performed in the opposite order, *b* followed by *a*, the combined effect is a rotation through angle *b* + *a*. But addition of angles in the plane is commutative, so *a* + *b* = *b* + *a*. The additive notation conceals the key point: We are here composing *transformations.* If *A* is the transformation "through angle *a*" and *B* is the transformation "rotate through angle *b*," then we can write "first do *A*, then *B*" as *AB*, and "first do *B*, then *A*" as *BA*. In this case, of course, *AB* = *BA*: The "multiplication" of the two transformations is commutative.

So why all the fuss? Because in three (or more) dimensions, *none of this works anymore.* Suppose the object is the Earth. Let *A* be a rotation through 90° (from west to east as viewed from the front) about an axis joining the north and south poles; let *B* be a rotation through 90° (from south to north as viewed from the front) about an axis joining two points on the equator at longitudes ±90°. Then you can easily convince yourself, using a tennis ball or something similar, that now *AB* is not the same as *BA*. Think about where the north pole goes, for instance (Figure 47). Why the difference? In two dimensions, all rotations have the same axis — the fixed direction around which everything else spins. In this case, the axis of rotation points vertically out of the plane. In three or more dimensions, though, rotations can have many different axes.

Hamilton had been struggling for years to devise an effective algebra for three dimensions.

Eventually he found one, a number system that he called *quaternions.* But it was really an algebra of four dimensions, not three, and its multiplication was *non*commutative. He described his sudden insight in these words (his italics):

Quaternions ... started into life, fully grown, on the 16th of October, 1843, as I was walking with Lady Hamilton to Dublin, and came up to Brougham Bridge. That is to say, I *then and there* felt the galvanic circuit of thought closed, and the sparks which fell from it were the fundamental equations between *I, J, K; exactly such* as I have used them ever since. I pulled out, on the spot, a pocketbook, which still exists, and made an entry, on which, *at the very moment,* I felt that it might be worth my while to expend the labour of at least ten (or it might be fifteen) years to come. I felt a *problem* to have been at that moment *solved,* an intellectual *want relieved,* which had haunted me for at least *fifteen years* before.

In *Men of Mathematics,* Eric Temple Bell says that Hamilton immediately carved the equations in the

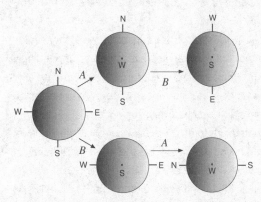

Figure 47 Rotations in three dimensions need not commute.

stone of the bridge, but this seems to be an invention.

Quaternions are remarkably similar to complex numbers, but now, instead of one "new" number i, there are three: i, j, k. These are the I, J, K of Hamilton's description in today's notation. A quaternion is a combination of these — for example

$$5 + 4i + 3j + 2k$$

Just as the complex numbers are two-dimensional, built from the two independent quantities 1 and i, so the quaternions are four-dimensional, built from the four independent quantities 1, i, j, and k. Addition of quaternions is what you would expect, and multiplication is governed by a small list of rules:

$$i^2 = j^2 = k^2 = -1$$

$$ij = k, \ ji = -k, \ jk = i, \ kj = -i, \ ki = j, ik = -j$$

Note that, for example, $ij \neq ji$; multiplication does not commute. These multiplication rules have a natural interpretation in which i, j, k are rotations about the three axes of a three-dimensional space, but in order to make the algebra work, a fourth, *independent* "axis" is needed to represent the ordinary number 1.

The most surprising consequence of these rules is that not only can you add, subtract, and multiply quaternions to get a quaternion, but you can also *divide* a quaternion by any (nonzero) quaternion and still get a quaternion. In modern language, the quaternions form a *division algebra*. Hamilton thought that his quaternions were the key to mathematical physics, a discovery at least as important as calculus. His friend Peter Guthrie Tait (1831–1901) agreed and wrote many articles urging the use of quaternions in physics. Cayley disagreed profoundly, and the physicists ignored them all and carried on using ordinary Cartesian coordinates.

In 1845 Cayley discovered an extension of the quaternions with eight independent basic numbers, that is, eight dimensions: these *Cayley numbers* (or *octonions*) threw away a further property of multiplication, the *associative law* $(ab)c = a(bc)$. The division algebra property remains valid, however. The dimensions of the real numbers, complex numbers, quaternions, and Cayley numbers are 1, 2, 4, 8, respectively. What comes next? The obvious guess is 16 — but is there a 16-dimensional division algebra? If there was, nobody could find it. If there wasn't, nobody could explain why.

As Many Dimensions As You Like

Ironically, mathematicians had been aware for some time that "spaces" of high dimensions arise entirely naturally, and have sensible physical interpretations, if the basic elements of space are considered to be different from points. Space, for instance, is three-dimensional in the sense that three numbers specify any *point,* but as Julius Plücker (1801–1868) pointed out in 1846, it takes four numbers to specify a *line* in space: two to determine where it hits some fixed plane and two more to determine its direction relative to that plane (Figure 48). Considered as a space of *lines,* then, our familiar space already has four dimensions, not three! Another such example would have delighted Abbott. Suppose we consider the plane to be composed not of its constituent points but

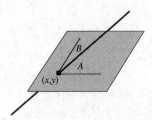

Figure 48 It takes four numbers to specify a line in space.

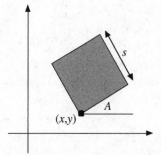

Figure 49 It takes four numbers to specify a square in space.

of all possible squares that can be drawn in it. Now what is its dimensionality? Well, it takes two numbers to specify where some particular corner of the square is, one more to say at what angle the square is tilted, and a fourth to tell us how big it is (Figure 49). Considered as "square space," the space of all possible squares, the plane is four-dimensional!

Of course, examples like this failed to convince the skeptics, because everything was still just a disguised description of good old two- or three-dimensional space. But imaginative examples, however evocative, *never* convince the skeptics, for their minds are closed. It never dawned on them that mathematics had been through all this before — that the same tired objections had been raised to "prove" that complex numbers and non-Euclidean geometry didn't exist and therefore couldn't possibly be of any use. But the whole art of mathematics is to make use of things that don't exist. Four cats exist, and so do four flowers — but *four*? You won't find a *four* in your front garden. "Four" is an abstract construct, not a thing. So is four-dimensional space — *and so is the mathematician's three-dimensional space.* Slowly, the more perceptive mathematicians, philosophers, and scientists

were coming to understand this, but there was still some way to go.

While Hamilton was trying to sell quaternions to the world, a high-school drop-out named Hermann Günther Grassmann (1809–1877), who had become a mathematics teacher, was busily inventing a much more general extension of the number system, an extension into space of any number of dimensions. He had the idea before Hamilton but published it a year later, in 1844, as *Die Lineale Ausdehnungslehre* (*Lectures on Lineal Extension*). His presentation was mixed up with mysticism, and his viewpoint was rather abstract, so his work remained obscure. In 1862 he issued a revised version, *Die Ausdehnungslehre* (*Lectures on Extension*, often translated as *The Calculus of Extension*), which was intended to be more comprehensible ... but it wasn't. Instead of four basic "numbers" 1, i, j, k, Grassmann used n of them:

$$e_1, e_2, ..., e_n$$

Combinations of these, such as $5e_1 + 6e_2 + ... + 99e_n$, he called *hypernumbers*. And he introduced two different kinds of multiplication. The *inner product* (symbolized by $\sqrt{}$) is defined by making $e_p \sqrt{e_q} = 1$ if $p = q$, and 0 otherwise; it is commutative, and it always produces an ordinary number. The *outer product* (symbolized by [ ]) satisfies

$$[e_p e_q] = -[e_q e_p] \text{ and } [e_p e_p] = 0$$

and in general it leads to quantities that lie outside the realm of hypernumbers.

Meanwhile, the physicists were heading along the same track but with an entirely different means of propulsion. Their main motivation was Maxwell's Equations for electromagnetism, in which both the electric and magnetic fields are vectors. *Vector* means that these quantities are not just magnitudes but magnitudes that have a *direction* in three-dimensional space: arrows, if you wish, aligned with the field, whose length shows how strong the field is, and whose direction shows which way it points. The arrow can be split into three separate *components,* aligned with the three axes. Maxwell's Equations (in the notation of the time) were eight in number, but they included two groups of three equations, one for each separate component of the electric (or magnetic, respectively) field. It would make life much easier for everyone if a formalism could be devised that would collect each such triple into a single vector equation. Maxwell managed to do this using quaternions, but only by separating out the vector part (the terms in i, j, k from the scalar part (the terms involving just 1).

Independently, the physicist Josiah Willard Gibbs (1839–1903) and the engineer Oliver Heaviside (1850–1925) found a way to stop trying to torture mathematical physics into the quaternionic strait-jacket. In 1881 Gibbs printed a private pamphlet, *Elements of Vector Analysis,* for his students. He explained that his vectors were similar to quaternions but had been developed for convenience in use rather than mathematical elegance. The material was written up by E. B. Wilson, and a joint book (*Vector Analysis*) was published in 1901. Heaviside came up with the same general ideas in the first volume of his *Electromagnetic Theory* (1893, 1899, 1912). To both Gibbs and Heaviside, as to Maxwell, a vector was a quaternion without any terms involving the ordinary number 1; that is, it is an expression

$$xi + yj + zk$$

where x, y, z are real numbers. Geometrically, this expression represents an "arrow," or directed line segment, running from the origin $(0, 0, 0)$ of three-dimensional space to the point with coordinates (x, y, z) (Figure 50). Its length is $\sqrt{x^2 + y^2 + z^2}$. The notation is a thinly disguised version of Grassmann's; we might just as reasonably write the vector as

$$xe_1 + ye_2 + ze_3$$

Thus all the different systems of Hamiltonian quaternions, Grassmannian hypercomplex numbers, and Gibbsian vectors converged on the same mathematical description: What really matters is the triple (x, y, z). After 250 years, the world's mathematicians and physicists had worked their way right back to Descartes, only now the coordinate notation was laden with endless riches of interpretation. Triples did not just represent points: You could do algebra with them.

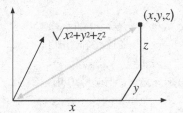

$\sqrt{x^2+y^2+z^2}$

(x,y,z)

z

y

x

Figure 50 A vector in three-dimensional space.

The mathematicians took one message from this work, the physicists another. The objective in physics (and engineering) was to develop a formal calculus of vectors in three-dimensional space and use it to solve the problems of physics. The mathematicians thought these new hypercomplex number systems were rather fun and wondered just how many variations on the theme there might be. To them, the question was not "Are they useful?" It was "Are they interesting?" It never occurred to the physicists that, for example, the two triples of Maxwell's Equations that referred to the electric and magnetic fields implied that the *electromagnetic* field was six-dimensional. And it never occurred to most mathematicians that the electromagnetic field was worth thinking about, anyway. So for several decades, the two went their separate ways — with one glorious exception, as we will shortly see.

The mathematicians focused mainly on algebra, on systems of *n* hypercomplex numbers. These were, in fact, *n*-dimensional spaces *plus* algebraic operations, but to begin with everyone thought algebraically, and geometric aspects were played down. One triumph of this style of algebraic rea-

soning was the startling discovery that the sequence 1, 2, 4, 8 — the dimensions of the real numbers, complex numbers, quaternions, and Cayley numbers, respectively — stops at that point. There is no 16-dimensional system of hypercomplex numbers that forms a division algebra. Obvious numerical patterns do not always extend. Numerological analogies do not always pan out.

Curved Space

The geometers responded to the algebraists' invasion of their territory. The key figure here is Riemann, a student of Gauss. Riemann was working for his Habilitation, which gave him the right to teach as a Privatdozent (a lecturer who could charge fees to the students). Earning the Habilitation, which was and still is a bit more advanced than a Ph.D., involves giving a special lecture on the candidate's own research. Following the usual procedure in such matters, Gauss asked Riemann to propose a number of topics, from which Gauss would make the final choice; the lecture was to be given in 1854. One of Riemann's proposals was "On the Hypotheses Which Lie at the Foundation of Geometry." As we saw, Gauss had done some marvelous work on the curvature of surfaces, notably his Most Excellent Theorem, so when Riemann offered to lecture on geometry, Gauss pounced. Riemann was terrified — he disliked public speaking anyway, and he hadn't fully worked out his ideas when he proposed the topic. However, he did know roughly what he had in mind, and it was explosive: a geometry of *n* dimensions, by which he meant systems of n coordinates $(x_1, x_2, ..., x_n)$ equipped with a notion of the distance between nearby points. He called such a space a *manifold* ("many-foldedness"). This proposal was radical enough, but there was another, even more brilliant feature: manifolds could be *curved*.

Riemann's intention was to find a formula for their curvature, one that would extend Gauss's own formula for $n = 2$. The formula was to be intrinsic to the manifold. That is, it would not make explicit use of any larger, perhaps simpler space in which the manifold might be embedded. Riemann's efforts to develop the notion of curvature in a space of n dimensions led him to the brink of a nervous breakdown. What made matters worse was that at the same time, he was helping Gauss's colleague Weber, who was trying to understand electricity. Riemann battled on despite the difficulties, and his efforts were rewarded. The interplay between electrical and magnetic forces led him to a new concept of "force," based on geometry. He had the same insight that led Einstein to General Relativity decades later: that forces can be replaced by curvature of space. In traditional mechanics, bodies travel along straight lines unless diverted by a force. In curved geometries, straight lines need not exist and paths are necessarily curved. Conclusion: If space is curved, what you experience when you are obliged to deviate from a straight line feels like a force. Now Riemann had the insight he needed to develop his lecture, which was a triumph. The word spread, accompanied by growing excitement. Soon scientists were giving popular lectures on the new geometry. Among them was Helmholtz, who gave many talks about beings that lived on a sphere or some other curved surface.

In 1870 William Kingdon Clifford (1845–1879) took up Riemann's suggestion that forces are related to the curvature of space, and in his *Space Theory of Matter,* he proposed an astonishing reinterpretation not just of space, but of matter as well.

I hold in fact: (1) That small portions of space are of a nature analogous to little hills on a surface which is on average flat. (2) That this property of being curved or distorted is continually passed from one portion of space to another after the manner of a wave. (3) That this variation in the curvature of space is really what happens in that phenomenon which we call the motion of matter, whether ponderable or ethereal [that is, particles or radiation]. (4) That in this physical world nothing else takes place but this variation, subject, possibly, to the law of continuity.

Clifford here anticipates not just Einstein's General Relativity but also some central features of quantum field theory. He even developed a four-dimensional model of electromagnetism based on Riemann's ideas. The technical aspects of Riemann's geometry of manifolds, now called differential geometry, were further developed by Eugenio Beltrami (1835–1900), Elwin Bruno Christoffel (1829–1900), and the Italian school under Ricci and Levi-Civita.

The algebraists had also been busy, developing computational techniques for n-variable algebra, the formal symbolism of n-dimensional space. One of these was the algebra of matrices (rectangular arrays of numbers) introduced by Cayley in 1855. Matrix algebra made it possible to calculate what things *did* in n-dimensional space, even if you didn't understand what any of it *meant*. What particularly interested Cayley was the existence of "invariants" — algebraic expressions that remained unchanged when their variables were transformed in certain ways. As a simple example, x^2 stays the same when x is replaced by $-x$. However, the transformations that interested Cayley were not that simple. The phenomenon of invariance had been discovered by George Boole in 1841, and by the 1850s Cayley and Sylvester were producing numerous papers on the topic. George Salmon (1819–1904) joined in, and the three became known as the "invariant trinity." The study of invariants led naturally into many algebraic equations in many unknowns — multidimensional geometry in all but name.

Gradually, as the new ideas took hold, an elegant, self-consistent geometric language for n-dimensional space came into being, supported by a formal algebraic computational system. You could *prove* it was self-consistent by using matrix algebra, so there was no harm in using geometric images. You just tried not to wonder what a 7-dimensional hyperplane in 11-dimensional space really meant! The geometric imagery even made it easier to prove theorems, and this, as far as the mathematicians were concerned, clinched the argument. Multidimensional "geometry" had proved its technical worth, so why not just accept it as an extension of conventional geometry and stop worrying? After

all, mathematics had swallowed distinctly less palatable things in the past and had thrived on the resulting intellectual diet.

The critics countered that the theorems of these newfangled "geometries" were all very well but that they weren't *interesting* because they referred to spaces that didn't exist. The algebraists fought back by pointing out that the algebra of n variables most certainly did exist and that anything that helped advance many different areas of mathematics must surely be interesting. Salmon wrote,

I have already completely discussed this problem [solving a certain system of equations] when we are given three equations in three variables. The question now before us may be stated as the corresponding problem in space of p dimensions. But we consider it as a purely algebraical question, apart from any geometrical considerations. We shall however retain a little of geometrical *language* … because we can thus more readily see how to apply to a system of p equations, processes analogous to those which we have employed in a system of three.

Do Higher Dimensions Exist?

The issue was brought to a head by Sylvester in 1869, who called (in the language of the time) for multidimensional geometry to come out of the closet and stop pretending it was just algebra in disguise. In a famous address to the British Association, later reprinted as "A Plea for the Mathematician," he pointed out that generalization is an important way to advance mathematics; that what matters is what is *conceivable,* not what corresponds in a simple-minded manner to physical experience; and that with a little practice, it is perfectly possible to visualize four dimensions (so four-dimensional space *is* conceivable). Incensed, the Shakespearean scholar Clement Mansfield Ingleby (1823–1886) invoked Kant to prove that three-dimensionality was an essential feature of space. (Ingleby did a lot of that kind of thing.) We now see that the nature of real space is irrelevant to the mathematical issues, but for a time, most British mathematicians agreed with Ingleby. Sylvester had been a little too imaginative for them. Some of the continental mathematicians were less hidebound. Here is Grassmann:

The theorems of the Calculus of Extension are not merely translations of geometrical results into an abstract language; they have a much more general significance, for while the ordinary geometry remains bound to three dimensions of [physical] space, the abstract science is free of this limitation.

That is, if multidimensional spaces do not exist, we'll just have to invent them anyway. They're much too useful to ignore.

Despite all the fuss, it was inevitable that eventually the new imaginative viewpoint would win out, provided that it was useful. And not in nuts-and-bolts terms, but in intellectual terms. Once multidimensional geometry began to impinge on the mainstream of mathematics and add power to the mathematician's elbow — once it helped mathematicians solve problems in areas different from multidimensional geometry itself — the new view-

point was never going to go away. The Old Guard would die out (literally); the next generation wouldn't have the same hang-ups and would find it incredible that there had ever been anything to disagree about. Sylvester saw this. In "A Plea for the Mathematician," he said,

There are many who regard the alleged notion of a generalized space as only a disguised form of algebraic formulization; but the same might be said with equal truth of our notion of infinity, or of impossible lines, or lines making a zero angle in geometry, the utility of dealing with which no one will be found to dispute. Dr. Salmon in his extension of Chasles' theory of characteristics to surfaces, Mr. Clifford in a question of probability, and myself in the theory of partitions, and also in my paper on barycentric projection, have all felt and given evidence on the practical utility of handling space of four dimensions as if it were conceivable space.

It is worth asking what the skeptics thought they were objecting *to*. If, as they claimed, four-dimensional geometry was merely an intellectual viewpoint — a reformulation in "geometric" language of valid algebraic calculations — what possible harm could it do? No one was proposing to reengineer the universe into a new dimension. But in any case, they were behind the times: The terms of the debate were becoming increasingly misplaced. Who *cares* how many dimensions physical space has? We're talking mathematics, and that's a different ball game entirely.

By 1900 mathematics was beginning to move toward a radical new conception of its own foundations. No longer did it have to be rooted in physical reality ("real" numbers, geometry of "real" space). As long as it was logically consistent, anything went. An influential figure in this change of underlying philosophy was Hilbert. Having improved on Euclid's axioms for geometry, Hilbert was after even bigger game: to axiomatize the whole of mathematics. And the more he dug into such questions, the more he came to realize that it is the logical structure of mathematics that counts and that logic is much more dependent on the relations between things than on what those things "really" are. It followed that the question "Do higher dimensions exist?" was meaningless until one decided what one *meant* by mathematical existence. And whatever that is (which is still a major debating point), it has nothing to do with physical reality.

Forget whether the fourth *dimension* exists: Does the number *four* exist?

Also around 1900, Sylvester's predictions were coming true. There was an explosion of mathematical and physical areas in which concepts of multidimensional space were making a serious impact. One was relativity, which is best considered a special kind of four-dimensional space-time geometry. As formulated by Minkowski in 1908, the three spatial coordinates (x, y, z), together with time t, form a four-dimensional *space-time* with coordinates (x, y, z, t). Any such "point" is called an *event*: It is like a point particle that winks into existence at just one moment in time and then winks out again. Relativity is really about the physics of events. In traditional mechanics, a particle moving through space occupies coordinates $(x(t), y(t), z(t))$ at time t, and this position changes as time passes. From Minkowski's space-time viewpoint, the collection of all such points is a curve in space-time, the *world line* of the particle. The most important new ingredient in relativity is that the appropriate measure of "distance" is not

$$\sqrt{x^2 + y^2 + z^2 + t^2}$$

as it would be in standard four-dimensional Euclidean space, but

$$\sqrt{x^2 + y^2 + z^2 - c^2t^2}$$

where c is the speed of light. This quantity, which is called the *interval,* emerges from Maxwell's equations of electromagnetism. For physically realistic dynamics, the quantity inside the square root must be positive; otherwise, the interval would be imaginary. Equivalently, no particle can travel faster than light. And associated with any event is its *light cone,* which determines its future (the events that are accessible from it by a moving particle) and its past (the events from which it is accessible by a moving particle). Every particle's world line must always stay within the appropriate light cone. This geometric imagery (Figure 51) makes relativity much easier to understand than it is if you think purely in terms of algebraic formulas, and you don't have to be *very* imaginative to concede that space-time exists. Given that, it obviously must have four dimensions: north-south, east-west, up-down, and pastfuture. The geometry involved, though, is very different from the Newtonian picture of the past, present, and future. The incorporation of gravity, achieved in General Relativity, made heavy use of

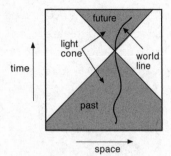

Figure 51 Light cones and world lines in Minkowski's relativistic geometry. Here space is schematically shown as being one-dimensional.

Riemann's revolutionary geometries. Einstein's key principle here was one of symmetry: The laws of physics should be the same at every point in space and at every instant of time. The implications of this principle were determined by using the theory of differential invariants, an extension of the algebraic invariant theory of Boole, Cayley, and Sylvester.

In relativity, though, "the" fourth dimension has a single, fixed interpretation: time. Mathematicians preferred a more flexible notion of dimensionality and "space," and as the late nineteenth century flowed into the early twentieth, mathematics itself seemed increasingly to demand acceptance of multidimensional geometry. The theory of functions of two complex variables, a natural extension of complex analysis, required thinking about space of two complex dimensions — but each complex dimension boils down to two real ones, so like it or not, you're looking at a four-dimensional space. Riemann's manifolds and the algebra of many variables provided further motivation. Yet another stimulus was Hamilton's 1835 reformulation of me-

chanics in terms of "generalized coordinates," a development initiated by Lagrange in his *Mécanique Analytique* (*Analytical Mechanics*) of 1788. A mechanical system has as many of these coordinates as it has "degrees of freedom" — that is, ways to change its state. In fact, *number of degrees of freedom* is just *dimension* in disguise. For example, it takes six generalized coordinates to specify the configuration of a (rudimentary) bicycle: one for the angle at which the handlebars sit relative to the frame, one each for the angular positions of the two wheels, another for the axle of the pedals, and two more for the rotational positions of the pedals themselves. A bicycle is, of course, a three-dimensional object, but the space of possible configurations of the bicycle is *six*-dimensional, which is one of the reasons why learning to ride a bicycle is hard until you get the knack. Your brain has to construct an internal representation of how those six variables interact — you have to learn to navigate in the six-dimensional geometry of bicycle space. For a moving bicycle, there are six corresponding velocities to worry about too: The dynamics is, in essence, *twelve*-dimensional.

By 1920 this concurrence of physics, mathematics, and mechanics had carried the day, and the use of geometric language for many-variable problems — multidimensional geometry — had ceased to raise eyebrows, except perhaps among philosophers. By 1950 the mathematicians' natural tendency was to formulate everything in *n* dimensions from the beginning. Limiting your theories to two or three dimensions seemed old-fashioned and ridiculously confining. The language of higher-dimensional space permeated every area of science and even invaded such subjects as economics and genetics. Today's virologists, for instance, think of viruses as "points" in a space of DNA sequences that could easily have several hundred dimensions. What they mean by this, fundamentally, is that the genomes of these viruses are several hundred DNA bases long, but the geometric image goes beyond mere metaphor. It provides an effective way to *think* about the problem. Let's see how.

An organism's DNA can be represented as a sequence of four letters, A, C, G, and T, which are the first letters in the names of certain molecular components called *bases*. The DNA of a virus might look something like this:

CCATGGAC ... TTTAGC

with several hundred letters in total. In the "space" imagery, this sequence is thought of as a list of several hundred coordinates. In conventional geometry, each coordinate is a number, but here each specific coordinate is one of only four possibilities: A, C, G, or T. We can think of this setup as being analogous to a hypercube, where the coordinates of the corners are restricted to the possibilities 0 or 1, and each possible sequence of 0s and 1s corresponds to a unique corner. A hypercube is like DNA with only two choices for bases; real DNA has four. The possible virus DNA sequences can therefore be visualized (conceptually, not pictorially) as forming some kind of "supercube," and any given virus's DNA corresponds to some "corner" of this supercube. The "edges" of the supercube connect neighboring corners, and two corners are neighbors if the corresponding sequences differ in exactly one position. That is, an edge corresponds to a *point mutation* of the DNA — a change in a single base. Note how the natural mathematical structure parallels basic biological features. The final step in setting up the spatial structure is to introduce a notion of distance, and here we abandon Pythagoras in favor of an idea that originated in information theory. We define each edge of the supercube to be one unit long, and the distance between any two corners is equal to the total number of edges in the shortest path that connects them. With this notion of distance, "spheres" are shaped more like cubes, but the familiar language is useful despite such differences.

To illustrate the value of geometric imagery, let's see whether we can gain some insight into the enormous variability of viruses. They mutate very frequently, and most viruses occur in a wide variety of different forms. Where does all this variety come from? We've just seen that a single mutation in a virus's DNA leads to a neighboring sequence in the supercube, so with just one point mutation, a virus can change into any of its potential neighbors. Therefore, each virus's DNA sequence sits at the center of a "sphere" of unit radius, and everything else inside that sphere corresponds to a neighboring DNA sequence. In order to understand the enormous potential for variability in virus DNA, all we need to observe is that such a sphere is very large, even though its radius is unity. The "hypervolume" of a hypersphere grows extremely rapidly as the number of dimensions of the space increases. For example, in a virus with a DNA sequence 200 bases long, which is relatively short, the sphere of neighbors of radius 10, say — that is, up to ten point mutations — contains approximately 10^{23} sequences, which is gigantic.

There are other ways to understand this fact mathematically, but to anyone with experience of multi-dimensional spaces, the reasoning just outlined comes very naturally. Of course, there is more to a virus than just a DNA sequence: Many sequences do not actually correspond to viable viruses. Hence the geometric reasoning needs to be more subtle than our simple example. Nonetheless, this example gives a good indication of how multidimensional geometry can enhance scientific understanding. The imagery of higher-dimensional spaces has become so commonplace that its users are seldom conscious of how strange such concepts used to be. Space of four dimensions — indeed of hundreds, or millions — is now entirely respectable in mathematics and science.

However, this does not mean that the spirit world exists, that ghosts now have a credible home, or that one day we might receive a visit from the Hypersphere, a creature from the fourth dimension who would manifest himself to us as a sphere whose size kept mysteriously changing and who could shrink to a point and vanish from our universe. However, physicists working in the theory of superstrings currently think that our universe may actually have *ten* dimensions, not four.

Wild as that speculation might seem, it is tame compared to some of the ideas now floating around in fundamental physics. Superstrings have more exotic cousins called *branes*. The name comes from *membrane* and leads to standard physicists' jokes such as using "*p*-brane" for a *p*-dimensional brane. A brane is like a string but has more dimensions. Our entire universe may be just a low-dimensional brane in a much higher-dimensional multiverse or megaverse — just like two-dimensional Flatland, sitting in a surrounding space of three dimensions. The Big Bang may have been triggered when a sedate, unchanging, rather dull brane was rammed by another brane that emerged from the hidden dimensions. This caused the dull brane to explode into frenetic activity, one consequence of which was us. We and everything around us are vibrations in a multidimensional space, originally set in motion when a brane that was traveling too fast rear-ended a stationary brane.

The mathematics behind these speculations is deep, beautiful, and important, even if it turns out that the physics is wrong. But if the physics is right, we are on the verge of a revolution in our understanding of the universe that will trump both relativity and quantum mechanics. Indeed, it will unify both of these remarkable physical theories into a single, coherent whole and will provide a conceptually elegant explanation of space, time, matter, and the origins of the universe. The potential importance of high-dimensional geometry for real physical space and time has suddenly become immense.

This is all very well, but if the geometry of space-time is so radically different from what we observe in our daily lives, why don't we notice? Right now physicists think that we've never noticed the extra six dimensions because they are curled up too tightly for us to detect them — a revival of C. H. Hinton's explanation of why the fourth dimension does not make itself manifest. However, they are beginning to consider another possibility: The other six dimensions (and it may not even be six) are actually quite big, but we float inside a four-dimensional brane inside the ten-dimensional space, and the laws of physics prevent us from moving in those six "hidden" directions.

Just like A. Square in his Flatland plane.

Creatures from the fourth dimension? Arrant nonsense.

It's creatures from the *tenth* dimension that we need to worry about!

BIBLIOGRAPHY OF
EDWIN ABBOTT ABBOTT

This bibliography was compiled from the *National Union Catalog, pre-1956 Imprints*, Mansell, London 1968 (all extant Library of Congress index cards and titles); the *British Library General Catalogue of Printed Books*; the *English Catalogue of Books*, Marston Low, London; and original copies in the archives of the City of London School. Reprints made without essential changes are listed sequentially by date. New editions are in square brackets.

Books

A Shakespearian Grammar: An Attempt to Illustrate Some of the Differences Between Elizabethan and Modern English. For the Use of Schools. Macmillan, London 1869. [2nd ed. Macmillan, London 1870, 1871, 1872, 1873, 1874, 1875. 3rd ed. Macmillan, London, 1876, 1877, 1878, 1879, 1881,1883, 1884, 1886, 1888; Macmillan, London and New York 1891, 1894, 1897, 1901, 1905, 1909, 1913, 1919, 1925, 1929.]

Bible Lessons (Part I: Old Testament. Part II: New Testament). Macmillan, London and New York 1871,1872, 1875. [2nd ed. 1879. 3rd ed. 1890.]

[with J.R. Seeley] *English Lessons for English People.* Seeley, Jackson & Halliday, London 1871, 1875, 1880; Roberts Brothers, Boston 1872, 1873, 1874, 1876, 1878, 1880, 1881, 1883, 1884, 1888, 1891, 1893, 1898; Little, Brown, Boston 1901.

Good Voices: A Child's Guide to the Bible. Macmillan, London 1872.

How to Write Clearly: Rules and Exercises on English Composition. (10th ed. — previous editions by others.) Seeley, Jackson and Halliday, London 1872, 1874; Roberts Brothers, Boston 1876, 1879, 1880, 1882, 1883, 1884, 1885, 1886, 1887, 1888, 1890, 1891, 1892, 1893, 1894, 1895, 1896, 1898, 1899, 1900; Little, Brown, Boston 1907, 1912, 1914.

Latin Prose Through English Idiom: Rules and Exercises on Latin Prose Composition. Seeley, Jackson & Halliday, London 1873, 1877, 1884, 1896, 1899. [With additions by E. R. Humphreys, Allyn and Bacon, Boston 1876, 1878, 1882, 1886, 1888.]

Handbook of English Grammar. J. Martin, London 1873. [2nd ed., revised and augmented by William Moore, J. Martin, London 1874.]

Parables for Children, with Illustrations. Macmillan, London 1873.

Early Lessons in English Grammar. Seeley, London 1874.

How to Parse: An Attempt to Apply the Principles of Scholarship to English Grammar. With Appendices on Analysis, Spelling, and Punctuation. Roberts Brothers, Boston 1874, 1878, 1879, 1881, 1882, 1883,1889, 1892, 1897, 1898; Little, Brown, Boston 1900, 1908; Seeley, Jackson & Halliday, London 1875, 1880, 1885, 1902.

Cambridge Sermons: Preached Before the University. Macmillan, London 1875. [2nd ed. 1875.]

How to Tell the Parts of Speech, or, Easy Lessons in English Grammar. Seeley, Jackson & Halliday, London 1875. [2nd ed. *How to Tell the Parts of Speech,* an Introduction to English grammar, Seeley, Jackson & Halliday, London 1880, 1890; Seeley, Service, London 1919; American ed. revised and enlarged by John G. R. McElroy, Roberts Brothers, Boston 1881, 1883, 1885, 1887, 1892; Little, Brown, Boston 1900, 1919.]

Good Voices: A Child's Guide to the Bible and Parables for Children. Macmillan, London 1875.

Bacon and Essex: A Sketch of Bacon's Earlier Life. Seeley, Jackson & Halliday, London 1877.

Through Nature to Christ or The Ascent of Worship Through Illusion to the Truth. Macmillan, London 1877.

[Anon.] Philochristus: Memoirs of a Disciple of the Lord. Macmillan, London 1878; Roberts Brothers, Boston 1878. [2nd ed. Macmillan, London 1878; Roberts Brothers, Boston 1878. 3rd ed. Macmillan, London 1916, 1926.]

Oxford Sermons Preached Before the University. Macmillan, London 1879.

["by the author of Philochristus"] Onesimus: Memoirs of a Disciple of Paul. Macmillan, London 1882; Roberts Brothers, Boston 1882.

Hints on Home Teaching. Seeley, Jackson and Halliday, London 1883. [2nd ed. Seeley, Jackson and Halliday, London 1883; American Journal of Education, Hartford 1884.]

[with W. G. Rushbrooke] The Common Tradition of the Synoptic Gospels in the Text of the Revised Version. Macmillan, London 1884.

["A. Square"] Flatland: A Romance of Many Dimensions. Illustrated by the Author. Seeley, London 1884. [2nd ed. Seeley, London 1884; Roberts Brothers, Boston 1885, 1891; Little, Brown, Boston 1886, 1889, 1907, 1915. 3rd ed. with an introduction by William Garnett, B. Blackwell, Oxford 1926; Little, Brown, Boston 1926, 1927, 1928, 1929. 4th ed. B. Blackwell, Oxford 1932; Little, Brown, Boston 1937, 1941. 5th ed. B. Blackwell, Oxford 1944, 1950. 6th ed. B. Blackwell, Oxford 1950; Barnes & Noble, New York 1951. With introduction by Banesh Hoffman, Dover Publications, New York 1952, 1953. There are so many subsequent editions that the remainder, and numerous foreign translations, will not be listed.]

Francis Bacon: An Account of His Life and Works. Macmillan, London 1885.

Via Latina: A First Latin Book. Including Accidence, Rules of Syntax, Exercises, Vocabulary, and Rules for Construing. Seeley, London 1886, 1893, 1903, 1907.

["by the author of Philochristus and Onesimus"] The Kernel and the Husk: Letters on Spiritual Christianity. Macmillan, London 1886; Roberts Brothers, Boston 1887.

The Latin Gate: A First Latin Translation Book. Seeley, London 1889.

Newmanianism: A Preface to the Second Edition of Philomythus, Containing a Reply to the Editor of the "Spectator" [R. H. Hutton], a Few Words to Mr Wilfrid Ward, and Some Remarks on Mr. R. H. Hutton's "Cardinal Newman." Macmillan, London and New York 1891.

Philomythus: An Antidote Against Credulity (a Discussion of Cardinal Newman's Essay on Ecclesiastial Miracles). Macmillan, London 1891. [2nd ed. Macmillan, London 1891.]

The Anglican Career of Cardinal Newman, vol. 1. Macmillan, London and New York 1892.

The Anglican Career of Cardinal Newman, vol. 2. Macmillan, London and New York 1892.

Dux Latinus: A First Latin Construing Book. Seeley, London 1893.

The Spirit on the Waters: The Evolution of the Divine from the Human. Macmillan, London 1897.

St. Thomas of Canterbury: His Death and Miracles, vol. I. A. & C. Black, London 1898.

St. Thomas of Canterbury: His Death and Miracles, vol. II. A. & C. Black, London 1898.

Clue: A Guide Through Greek to Hebrew Scripture. A. & C. Black, London 1900. (Diatessarica Part I)

The Corrections of Mark: Adopted by Matthew and Luke. A. & C. Black, London 1901. (Diatessarica Part II)

Contrast; or, a Prophet and a Forger. A. & C. Black, London 1903. [Contents, Introductions I and II, and Appendix V of From Letter to Spirit.]

From Letter to Spirit: Attempt to Reach Through Varying Voices the Abiding Word. A. & C. Black 1903. *(Diatessarica: Part III)*

Paradosis; or "In the Night, in Which He Was (?) Betrayed." A. & C. Black, London 1904. *(Diatessarica: Part IV)*

Johannine Vocabulary. A Comparison of the Words of the Fourth Gospel with Those of the Three. A. & C. Black, London 1905. *(Diatessarica: Part V)*

Silanus the Christian. A. & C. Black, London 1906.

Johannine Grammar. A. & C. Black, London 1906. *(Diatessarica: Part VI)*

Apologia: An Explanation and Defence. A. & C. Black, London 1907.

Notes on New Testament Criticism. A. & C. Black, London 1907. *(Diatessarica: Part VII)*

Indices to Diatessarica: with a Specimen of Research. A. & C. Black, London 1907. [Compiled by Abbott's daughter Mary.]

The Message of the Son of Man. A. & C. Black, London 1909.

The Son of Man; or, Contributions to the Study of the Thoughts of Jesus. Cambridge University Press, Cambridge 1910. *(Diatessarica Part VIII)*

Light on the Gospel from an Ancient Poet. Cambridge University Press, Cambridge 1912. *(Diatessarica Part IX)*

The Fourfold Gospel. Section I: Introduction. Cambridge University Press, Cambridge 1913. *(Diatessarica Part X, Section I)*

Miscellanea Evangelica, vol. I. Cambridge University Press, Cambridge 1913.

Miscellanea Evangelica, vol. II: *Christ's Miracles of Feeding.* Cambridge University Press, Cambridge 1913, 1915.

The Fourfold Gospel. Section II: The Beginning. Cambridge University Press, Cambridge 1914. *(Diatessarica Part X, Section II)*

The Fourfold Gospel. Section III: The Proclamation of the New Kingdom. Cambridge University Press, Cambridge 1915. *(Diatessarica Part X, Section III)*

The Fourfold Gospel, Section IV: The Law of the New Kingdom. Cambridge University Press, Cambridge 1916. *(Diatessarica Part X, Section IV)*

The Fourfold Gospel, Section V: The Founding of the New Kingdom or Life Reached Through Death. Cambridge University Press, Cambridge 1917. *(Diatessarica: Part X, Section V, The Completion of the Work)*

Minor Works

"The Church and the Congregation." In Walther L. Clay (ed.), *Essays on Church Policy.* London 1868, 158–191.

The Proposed Examination of First Grade Schools by the Universities. Macmillan, London 1872.

"On Teaching the English Language." London College of Preceptors, *Lectures on Education* 1, 1872.

A Concordance to the Works of Alexander Pope. By Edwin Abbott [Abbott's father] with an introduction by Edwin A. Abbott. Chapman & Hall, London 1875; D. Appleton, New York 1875.

"Gospels." In *Encyclopaedia Britannica* (9th ed.) 1875.

On the Teaching of Latin. London Association of Schoolmistresses, 1881.

"Justin's Use of the Fourth Gospel." Reprinted from *The Modern Review,* July 1882, October 1882, London 1882.

Three Lectures on Subjects Connected with the Practice of Education. Delivered in the University of Cambridge in the Easter term 1882: H. W. Eve, "On Marking"; Arthur Sidgwick, "On Stimulus"; E.A. Abbott, "On the Teaching of Latin Verse Composition."

The Promus of Formularies and Elegancies (Being Private Notes circ. 1594 Hitherto Unpublished) by Francis Bacon, Illustrated and Elucidated by Passages from Shakespeare. By Mrs. Henry Pott [Constance Mary Pott, *née* Fearon] with preface by Edwin A. Abbott. Houghton Mifflin, Boston 1883.

"Genuineness of the Second Peter." *Southern Press Review,* April 1883.

Recent English Pedagogy. Hints on Home Training and Teaching. With lectures by Canon Farrar, Professors

Huxley, Quick, Laurie, and Meiklejohn, and the contents of other recent pedagogical treatises. *American Journal of Education*, Hartford 1884.

Bacon's Essays, 7th edition. By Francis Bacon, edited with an introduction by Edwin A. Abbott. Seeley, London 1889.

"On the Teaching of English Grammar." In *Teaching and Organisation*. Percy Arthur Barnett (ed.), London 1897, 98–135.

" 'Righteousness' in the Gospels." *Proceedings of the British Academy,* vol. 8 (1817–1918), 351–363. Reprinted separately, Oxford University Press, London 1918.

"Illusion in Religion." *Essays for the Times* 8. F. Griffiths, London.

"Teaching of English." Portion of some periodical, no. 103, vol. 18, 33–39.

BIBLIOGRAPHY OF CHARLES HOWARD HINTON

This bibliography was compiled from the *National Union Catalog, pre-1956 Imprints*, Mansell, London 1968 (all extant Library of Congress index cards and titles).

Books

Chapters on the Art of Thinking, and Other Essays. By James Hinton, edited by C.H. Hinton. K.C. Paul, London, 1879.

"What Is the Fourth Dimension?" Reprinted from the University magazine 1880. [2nd ed. From *The Cheltenham Ladies' College Magazine* no. VIII, September 1883, 31–52. 3rd ed. S. Sonnenschein, London 1897. 4th ed. S. Sonnenschein, London 1910.]

Science Notebook. John Haddon, London 1884. [No copies of this book are known to be extant.]

Scientific Romances. W. Swan Sonnenschein, London 1884–1885. [Binder's title: pamphlets.]

Scientific Romances, 1st series. Sonnenschein, London 1886. [Contents: What Is the Fourth Dimension?; The Persian King; A Plane World; A Picture of Our Universe; Casting Out the Self. 2nd ed. G. Allen & Unwin, London 1925.]

Scientific Romances, No. [1]–7. Swan Sonnenschein, Lowrey, London 1886, 2 vols. [Vol. 1 Contents: What Is the Fourth Dimension?; The Persian King; A Plane World; A Picture of Our Universe; Casting Out the Self. Vol 2. Contents: The Education of the Imagination; Many Dimensions.]

A New Era of Thought. S. Sonnenschein, London 1888. [2nd ed. with 230 pages instead of 216. S. Sonnenschein, London 1900. 3rd ed. Sonnenschein, London 1910 (imprint covered by label: This book is now published by G. Allen & Unwin, London)] "The MSS which formed the basis of this book were committed to us by the author.... It was his wish that we should construct upon them a much more complete treatise than we have effected.... Part I has been printed almost exactly as it came from his hands.... Part II has been written from a hurried sketch.... Chapter XI of this part [and Appendix E] have been entirely re-written by us.... Appendix H, and all the exercises have ... been written solely by us." From the preface, signed: Alicia Boole, H. John Falk.

Stella and An Unfinished Communication; Studies of the Unseen. S. Sonnenschein, London 1895; Macmillan, New York 1895.

Scientific Romances, 2nd series. S. Sonnenschein, London 1902. [Contents: The Education of the Imagination; Many Dimensions; Stella; An Unfinished Communication. 2nd ed. G. Allen & Unwin, London 1922.]

The Fourth Dimension. S. Sonnenschein, London 1904; J. Lane, New York 1904. [2nd ed. with 270 pages instead of 247. 3rd ed. 1912 G. Allen, London 1906. 4th ed. G. Allen & Unwin, London 1921. 5th ed. G. Allen & Unwin, London 1934. 6th ed. G. Allen & Unwin, London 1951.]

A Language of Space. Swan, Sonnenschein, London 1906.

An Episode of Flatland: or, How a Plane Folk Discovered the Third Dimension. To Which Is Added a History of Unaea. S. Sonnenschein, London 1907.

Minor Works

"A Mechanical Pitcher." *Harper's Weekly* (20 March 1897) 301–302.

"Review of Vivekananda's 'Yoga Philosophy.'" *Citizen* (May 1987) 62–64.

"The Recognition of the Fourth Dimension." Read before the Philosophical Society of Washington, 9 November 1901. The Society, Washington 1902. [*Philosophical Society of Washington, Bulletin* XIV (1902) 179–203.]

"The Oxford Spirit." *Independent* 54 (1902) 1217–1220.

"The Geometrical Meaning of Cayley's Formulae of Orthogonal Transformation." *Proceedings of the Royal Irish Academy* (29 November 1903) 59–65.

"The Fourth Dimension." *Harper's Monthly Magazine* (July 1904) 229–233.

"The Motion of a Baseball." *Yearbook of the Minneapolis Society of Engineers* (May 1908) 18–28.

SOURCES AND REFERENCES

Dictionary of National Biography (since 1917), Oxford University Press, Oxford; founded 1882 by George Smith.

English Catalogue of Books. Marston Low, London; Publishers Circular Ltd., London; Kraus Reprint Company, Millwood, New York 1976.

The National Union Catalog: pre-1956 Imprints. Mansell, London 1968.

www.theplays.org (searchable online version of the complete works of William Shakespeare).

Clifford W. Ashley. *The Ashley Book of Knots*. Doubleday, New York 1944.

Saint Augustine. *Confessions*, translated with an introduction by R.S. Pine-Coffin. Penguin, New York 1961.

Saint Augustine. *City of God*, translated by Henry Bettenson with an introduction by John O'Meara. Penguin, New York 1972.

Thomas F. Banchoff. "From *Flatland* to Hypergraphics: Interacting with Higher Dimensions." *Interdisciplinary Science Reviews* 1990. (www.geom.umn.edu/~banchoff/ISR.ISR.html)

Thomas F. Banchoff. "*Flatland*, A New Introduction." In *Flatland* (Edwin A. Abbott). Princeton University Press, Princeton 1991.

Thomas F. Banchoff. *Beyond the Third Dimension: Geometry, Computer Graphics, and Higher Dimensions*. Scientific American Library, Freeman, San Francisco 1995.

Stephen Baxter. "Wild Extravagant Theories: The Science of the Time Machine." *Picocon 13* (4 February 1996). Imperial College, London. (www.sam.math.ethz.ch/~pkeller/BAXTER/Articles/PicoconTalk.html)

James. E. Beichler. "The *Psi*-ence Fiction of H.G. Wells." *Yggdrasil* 1997. (ourworld.compuserve.com/homepages/Paraphys/psifi.htm)

Brian Butterworth. *The Mathematical Brain*. Macmillan, London 1999.

John Clute and Peter Nicholls. *The Encyclopaedia of Science Fiction*. Orbit, London 1993.

Marianne Colloms and Dick Weindling. *The Good Grave Guide to Hampstead Cemetery, Fortune Green*. Camden History Society, London 2000.

H.S.M. Coxeter. *Regular Polytopes*. Macmillan, New York 1948. 2nd ed. 1963.

Stanislas Dehaene. *The Number Sense*. Allen Lane, Harmondsworth 1997.

A.K. Dewdney. *Two-Dimensional Science and Technology* (duplicated notes). Ontario 1980.

Fyodor Dostoyevksy. *The Brothers Karamazov*. Penguin, Harmondsworth 1993. (Original 1880.)

A.E. Douglas-Smith. *City of London School* (2nd ed.). Blackwell, Oxford 1965.

Kenneth Falconer. *Fractal Geometry*. Wiley, New York 1990.

John Fauvel, Raymond Flood, and Robin Wilson (eds.). *Möbius and His Band*. Oxford University Press, Oxford 1993.

Christian Gottlieb. "The Simple and Straightforward Construction of the Regular 257-gon." *Mathematical Intelligencer* 21, no. 1 (1999) 31–37.

John Gribbin. *In Search of Schrödinger's Cat.* Wildwood House, London 1984.

Sir Thomas L. Heath (translator). *Euclid — The Thirteen Books of the Elements,* vols. 1–3. Dover Publications, New York 1956.

John G. Hocking and Gail S. Young. *Topology.* Addison-Wesley, Reading, Mass. 1961.

A. G. Howson. *A History of Mathematics Education in England.* Cambridge University Press, Cambridge 1982.

D. F. Lawden. *Elements of Relativity Theory.* Wiley, New York 1985.

Henry Parker Manning. *Geometry of Four Dimensions.* Macmillan, New York 1914.

P. D. Ouspensky. "The Fourth Dimension." In *A New Model of the Universe,* 1931. Reprint: Random House, New York 1971.

Abraham Pais. *"Subtle Is the Lord ..." The Science and Life of Albert Einstein.* Oxford University Press, Oxford 1982.

Joan L. Richards. *Mathematical Visions: The Pursuit of Geometry in Victorian England.* Academic Press, Boston 1988.

N. J. A. Sloane and Simon Plouffe. *The Encyclopedia of Integer Sequences.* Academic Press, San Diego 1995.

Alan D. Solomon. "Pick a Number: What Edwin Abbott Did Not Know About Flatland." *Oak Ridge National Laboratory Review* 25, no. 2, 1994.

D. M. Y. Somerville. *An Introduction to the Geometry of N Dimensions.* Dover Publications, New York 1958.

Daniel Stashower. *Teller of Tales: The Life of Arthur Conan Doyle.* Allen Lane, Harmondsworth 1999.

Jonathan Swift. *Gulliver's Travels and Other Writings* (ed. Louis A. Landa). Oxford University Press, Oxford 1976.

John L. Synge and Byron A. Griffith. *Principles of Mechanics* (3rd ed). McGraw-Hill, New York 1959.

H. G. Wells. *The Time Machine.* 1895. Reprinted in *Selected Short Stories,* H. G. Wells. Penguin, New York 1958.

Arthur Willink. *The World of the Unseen: An Essay on the Relation of Higher Space to Things Eternal.* Macmillan, New York 1893.

Amotz Zahavi, Avishag Zahavi, Amir Balaban (contributor), and Naama Zahavi-Ely. *The Handicap Principle: A Missing Piece of Darwin's Puzzle.* Oxford University Press, Oxford 1997.

J. C. F. Zöllner. *Transcendental Physics.* Beacon of Light Publishing, Boston 1901.

FURTHER READING

Eric Temple Bell. *The Development of Mathematics*. McGraw-Hill, New York 1945.

Elwyn R. Berlekamp, John H. Conway, and Richard K. Guy. *Winning Ways*. Academic Press, London 1982.

Dionys Burger. *Sphereland*. Apollo Editions, New York 1965.

A. K. Dewdney. *The Planiverse*. Poseidon Press, New York 1984.

A. K. Dewdney. *The Armchair Universe*. Freeman, New York 1988.

Michael J. Field and Martin Golubitsky. *Symmetry in Chaos*. Oxford University Press, Oxford 1992.

Martin Gardner. *The Unexpected Hanging and Other Diversions*. Simon & Schuster, New York 1969.

Martin Gardner. *Mathematical Carnival*. Simon & Schuster, New York 1969.

Martin Gardner. *The Ambidextrous Universe*. Penguin, New York 1970.

Reuben Hersh. *What is Mathematics, Really?* Jonathan Cape, London 1997.

Charles Howard Hinton (ed. Rudy Rucker). *Selected Writings of C. H. Hinton*. Dover Publications, New York 1980.

Michio Kaku. *Hyperspace*. Oxford University Press, Oxford 1994.

Morris Kline. *Mathematical Thought from Ancient to Modern Times*. Oxford University Press, Oxford 1972.

Desmond MacHale. *George Boole, His Life and Work*. Boole Press, Dublin 1985.

Heinz-Otto Peitgen and Peter H. Richter. *The Beauty of Fractals*. Springer-Verlag, New York 1986.

Terry Pratchett, Ian Stewart, and Jack Cohen. *The Science of Discworld*. Ebury Press, London 1999.

Rudy Rucker. *The Fourth Dimension and How to Get There*. Penguin, New York 1986.

Ian Stewart. *From Here to Infinity*. Oxford University Press, Oxford 1992.

Ian Stewart. *Visions Géométriques*. Belin, Paris 1994.

Ian Stewart. *Flatterland*. Perseus, Cambridge, Mass. 2001.

Jeffrey R. Weeks. *The Shape of Space*. Marcel Dekker, New York 1985.

Movies

Flatland. Director Eric Martin, 1965.

Flatland. Director Michele Emmer, 1982.

Flatland. Director Ladd Ehlinger Jr., Flatlands Productions Inc., 2004.

Flatland: The Movie. Director Jeffrey Travis, Flat World Productions 2007.

Edwin A. Abbott was born in London on December 20, 1838. Educated in St. John's College in Cambridge, he was ordained in 1862 and three years later was appointed headmaster of the City of London School, where he served until 1889. Abbott was a celebrated teacher, inspiring hundreds of students in a wide range of disciplines, from Sanskrit to chemistry. He was also a lifelong student of the classics and a writer of some renown. Although *Flatland* is certainly his most enduring work, Abbott wrote over fifty books, most of them scholarly works. He died in Hampstead on October 12, 1926.

Ian Stewart is Professor of Mathematics at the University of Warwick and Director of its Mathematics Awareness Centre. He is also a regular research visitor at the University of Houston, the Institute of Mathematics and Its Applications in Minneapolis, and the Santa Fe Institute. His many books include *From Here to Infinity*, *Nature's Numbers*, *Does God Play Dice?*, *The Problems of Mathematics*, and *Letters to a Young Mathematician*. His writing has appeared in *New Scientist*, *Discover*, *Scientific American*, and many newspapers in the U.K. and U.S. He lives in Warwick, England.